PHOTOGENERATED REAGENTS IN BIOCHEMISTRY AND MOLECULAR BIOLOGY

Hagan Bayley

Department of Biochemistry,
College of Physicians and Surgeons, Columbia University,
630 West 168 St., New York, NY 10032, USA

1983

ELSEVIER
AMSTERDAM · NEW YORK · OXFORD

ISBN – series: 0 7204 4200 1
– paperback: 0 444 80520 6
– hardbound: 0 444 80530 3

Published by:

ELSEVIER SCIENCE PUBLISHERS B.V.
P.O. BOX 211
1014 AG AMSTERDAM, THE NETHERLANDS

Sole distributors for the U.S.A. and Canada:

ELSEVIER SCIENCE PUBLISHING COMPANY, INC.
52 VANDERBILT AVENUE
NEW YORK, N.Y. 10017

Library of Congress Cataloging in Publication Data

Bayley, Hagan.
Photogenerated reagents in biochemistry and molecular biology.

(Laboratory techniques in biochemistry and molecular biology. 2nd ed. ; 12)
Includes index.
1. Biological chemistry--Laboratory manuals.
2. Molecular biology--Laboratory manuals. 3. Photochemistry--Laboratory manuals. I. Title. II. Series.
QP519.B36 1983 574.19'283 83-14191
ISBN 0-444-80530-3
ISBN 0-444-80520-6 (pbk.)

Printed in The Netherlands

LABORATORY TECHNIQUES IN BIOCHEMISTRY AND MOLECULAR BIOLOGY

Edited by

T.S. WORK – *Cowes, Isle of Wight (formerly N.I.M.R., Mill Hill, London)*
R.H. BURDON – *Department of Biochemistry, University of Glasgow*

ELSEVIER
AMSTERDAM · NEW YORK · OXFORD

To JRK and HGK

Dominus illuminatio mea

Preface

Fortunately, the experiments that can be done with photochemical reagents are too varied (and unpredictable) to allow the writing of a true laboratory manual. Instead I have tried to give an account of the possible experiments (Chapter 1), a description of the reagents that can be used (Chapter 2), and a discussion, rather than detailed protocols or dogmatic assertions, of how the experiments can be done (Chapters 3 to 6). In addition, the extensive bibliography of over 400 references will provide access to useful examples in the primary literature.

This monograph is an expansion and revision of the review I wrote earlier with my mentor Jeremy Knowles (Bayley and Knowles, 1977). Jeremy's earlier short review remains an excellent summary of the essential idea of photoaffinity labeling. Another review that will remain valuable is from Frank Westheimer's laboratory where photoaffinity labeling was invented (Chowdhry and Westheimer, 1977). Westheimer has also written a delightful short history and prospectus of the subject (Westheimer, 1980). Several additional reviews concerning special aspects of photochemical reagents are cited in this text.

I would like to thank my friends and colleagues who read the first draft, gave me many ideas for amendments, and told me of omissions and indeed a few errors. They were Josef Brunner, Mike Dockter, Boyd Haley, Arthur Karlin, John Katzenellenbogen, Koji Nakanishi, Fred Richards, Jim Staros and Bernadine Wisnieski. I also thank the many who sent preprints and reprints. I am indebted to the editors for their assistance and forbearance, particularly over the delay caused by my move to Columbia, to Marjorie Dunn who typed the first draft, to Olga Hanlon who produced the final version, and to all the others who helped me put together the manuscript.

I hope the reader will inform me of errors, omissions, or differences of opinion. I do regret not including a full chapter on crosslinking experiments with nucleic acids, and perhaps that can be repaired in the future.

HAGAN BAYLEY
Manhattan, New York, 1983

List of abbreviations

ACTH	corticotropin
AMP, ADP, ATP	adenosine 5′-mono-, di-, and triphosphates
atm	atmosphere (pressure)
cAMP, cGMP etc.	cyclic adenosine 3′:5′-monophosphate, etc.
CF_1	coupling factor 1, the hydrophilic portion of the proton-translocating ATPase of chloroplasts
CoA	coenzyme A
DEAE	diethylaminoethyl
DMF	dimethylformamide
DMSO	dimethylsulfoxide
DNA	deoxyribonucleic acid
dpm	decays per minute
DTT	dithiothreitol
EFG	elongation factor G
ϵ	molar extinction coefficient
Fab	monovalent immunoglobulin G fragment
GMP, GDP, GTP	guanosine 5′-mono-, di-, and triphosphates
HPLC	high pressure liquid chromotography
IEF	isoelectric focusing
IgG	immunoglobulin G
IR	infrared
K_d	dissociation constant
K_i	inhibition constant
K_m	Michaelis–Menten constant
λ_{max}	absorption maximum

Na,K-ATPase	sodium and potassium activated adenosine triphosphatase (Na pump)
NAD	nicotinamide adenosine dinucleotide
NEM	*N*-ethylmaleimide
N_3XMP etc.	Azidopurine nucleoside phosphates (8-azido derivatives unless otherwise indicated)
pUp	3′:5′-uridine diphosphate
ψ	quantum yield
RNA	ribonucleic acid
SDS	sodium dodecyl sulfate
THF	tetrahydrofluorine
TLC	thin layer chromatography
Tris	tris(hydroxymethyl)aminomethane
tRNA	transfer RNA
UV	ultraviolet

Contents

CHAPTER 1

The utility of photoaffinity labeling and related methods

Many biological molecules may be classified as ligands; others may be classified as receptors. For the purpose of this monograph it has proved convenient to extend the common usage of these terms (cf. Knowles, 1972). Here, ligands include enzyme substrates, allosteric effectors, haptens, neurotransmitters and hormones, and they bind to receptors, a term which embraces enzymes, immunoglobulins and the molecules which bind hormones, neurotransmitters and the like. Receptors are generally proteins. Ligands are more varied in structure and among them we find amino acids, sugars, nucleic acids, oligomers of these, the varied products of cellular metabolism, and foreign substances such as drugs.

Affinity labeling has been developed for investigating ligand–receptor interactions. The usual goal is either to identify a receptor in a mixture of candidates, which might be the polypeptides associated with a biological membrane or the subunits of a purified protein, or to locate one or more of the amino acid residues that make up the binding site of a receptor. In a chemical affinity labeling experiment (Singer, 1967; Shaw, 1970a,b; Glazer et al., 1975; Jakoby and Wilchek, 1977) the receptor is incubated with a modified ligand containing a functional group that, it is hoped, will react covalently with a residue at the binding site (Fig. 1.1). For instance, haloketones react with nucleophiles and one of the first successes of the affinity labeling technique was the identification of a histidine residue at the active site of chymotrypsin with *N*-tosyl-L-phenylalanine chloromethylketone (Schoellman and Shaw, 1963).

A useful chemical affinity reagent reacts more rapidly within the binding site than elsewhere because of its high 'local concentration' and, in the case of enzymes, because nucleophiles at the active site may be unusually

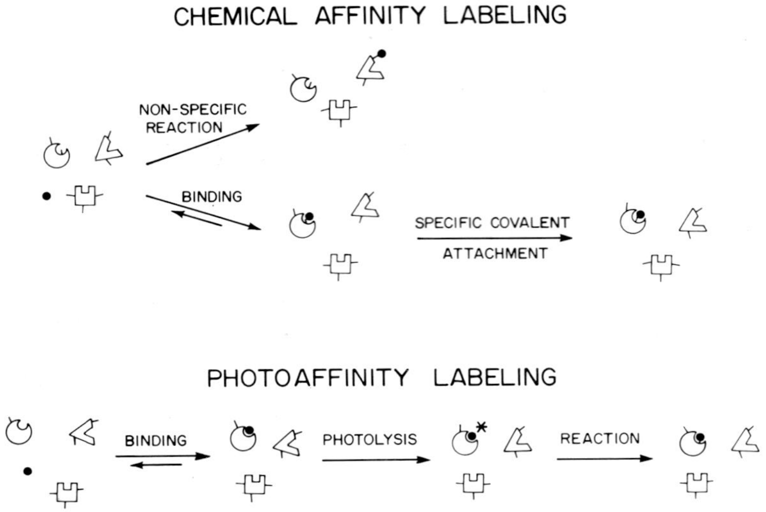

Fig. 1.1. Chemical affinity labeling and photoaffinity labeling. In a chemical affinity labeling experiment (*upper*) the ligand contains a reactive group designed to react with a functional group that might be present at the binding site of the receptor that recognizes the ligand. Under unfavorable conditions reaction may occur between the ligand and groups outside the binding site on the same protein or on a different molecule. For example, there might not be a suitably reactive or correctly oriented group at the binding site. In a photoaffinity labeling experiment *(lower),* the ligand contains a photoactivatable group. Reaction can be initiated after binding is complete, eliminating one source of non-specific labeling, and a particular functional group need not be present at the binding site as the most reactive photogenerated species can react even with carbon–hydrogen bonds. This is not to say that complications cannot arise (see Chapters 3 and 4).

reactive. The use of radiolabeled reagents greatly facilitates the identification of intact receptor subunits and derivatized fragments of them.

1.1. The advantages of photogenerated reagents

A photoaffinity reagent (Fig. 1.1.) is a ligand that is chemically inert but conceals a highly reactive intermediate that is unmasked by irradiation with near ultraviolet or visible radiation. The goals of photoaffinity labeling experiments are the same as those sought by chemical affinity labeling but the photochemical approach has several considerable advantages.

First, an inert ligand is desirable because it allows the investigator to perform important preliminary experiments with ease. Binding measurements and assays of biological activity are straightforward and unusual experiments that might not be feasible with a chemically reactive molecule can be done. For example, photolabile molecules may be used as haptens for antibody production and the hapten–immunoglobulin interaction subsequently explored (Fleet et al., 1969).

Second, the reaction of a photoaffinity reagent is initiated at will. In the case of kinetically inaccessible receptors such as those within living cells this is of crucial importance as chemical reagents may react with components of a biological preparation in a potentially misleading manner before they reach their target. The binding step has even been performed in living animals, and subsequently tissues have been removed and irradiated (see Section 4.2.). In favorable cases unbound photoaffinity reagents may be removed after the appropriate incubation period and the receptor–ligand complex rapidly irradiated before dissociation occurs, further reducing non-specific labeling. In areas related to photoaffinity labeling, bifunctional reagents containing both a chemically reactive group and a photochemically activatable group have been used in two-stage crosslinking experiments (see Chapter 5) and photochemical surface labeling reagents for membranes have been trapped inside cells before activation (see Chapter 6). In short, it is, in principle, less difficult to confine the reaction of a photogenerated reagent to the desired site.

A further advantage of photoaffinity reagents that reinforces the last conclusion is that intermediates formed by photolysis (see Chapter 2) are usually far more reactive than chemical reagents. Indeed, most of them are so reactive that it would be impossible to synthesize them and then add them to a receptor preparation. The most reactive photogenerated species can even attack functional groups such as carbon–hydrogen bonds which remain inert towards chemical reagents. In contrast, several different functional group-specific reagents may have to be tested before one suited to a particular problem is found. Extreme pH values, high temperatures or pre-reduction of disulfide bonds may be required for efficient reaction of a chemical reagent, and hydrophobic binding sites, largely containing hydrocarbon residues may prove to be impossible to derivatize. Misleading results can be obtained when attempts are made to label an inert binding site with a chemical reagent. The ideal photoaffinity reagent will not misbehave in this way.

Finally, these properties permit the use of photoaffinity reagents in experiments in which kinetic phenomena are examined, and it has proved possible to extend the resolution to the millisecond time-scale.

1.2. Some examples of photoaffinity labeling and related experiments

To induce the reader to proceed further, the remainder of this short chapter gives a brief account of some applications of photogenerated reagents. Five classes of photoaffinity labeling experiments are described followed by three examples of the related methods that are discussed in the later chapters of this book.

1.2.1. The identification of a receptor in a mixture of proteins

The insulin receptor is present in trace amounts in the plasma membrane of liver cells. Using radiolabeled arylazido derivatives of insulin, which on irradiation yield reactive nitrenes, components of the receptor with molecular weights of 135,000 and 90,000 daltons have been identified (Jacobs et al., 1979; Yip et al., 1980; Wisher et al., 1980). The identity of the two polypeptides has been confirmed by chemical crosslinking to radiolabeled insulin and by purification of the receptor by affinity chromatography.

In cases where it has proved impractical to assay for activity at each step, protein purification has been aided by tagging a fraction of the molecules with a photoaffinity reagent (e.g. the lactose carrier of *Escherichia coli*: Newman et al., 1981; and the β-adrenergic receptor of frog erythrocytes: Shorr et al., 1982).

1.2.2. The identification of a component of a multisubunit system

The binding of ligands to the β-adrenergic receptor of plasma membranes stimulates adenylate cyclase activity in a process that requires GTP. The existence of a separate GTP-binding protein (42,000 daltons), besides the hormone binding component and the cyclase, was confirmed by photoaffinity labeling with γ-(4-azidoanilino)-GTP (Pfeuffer, 1977). Recent purification of the GTP binding protein has confirmed the existence of a 42,000 dalton subunit that is the substrate for ADP-ribosylation by cholera toxin and NAD.

In another case, two ribosomal proteins at the peptidyl transferase site were derivatized with photolabile aminoacyl-tRNA analogs (Hsiung et al., 1974; Hsiung and Cantor, 1974).

1.2.3. The identification of a ligand binding site within a polypeptide

In favorable cases a labeled receptor may be degraded by proteolysis or chemical cleavage to identify the region of the polypeptide chain labeled by a photoaffinity reagent (Chapter 3). Occasionally the site of labeling has been precisely mapped. Kerlavage and Taylor (1980) were able to locate a single labeled tyrosine when the regulatory subunit of cAMP-dependent protein kinase II was labeled with 8-azido-cAMP.

1.2.4. The use of photoaffinity labeling in studies of function

Besides providing structural information, photoaffinity labeling is useful in studies of the function or cellular metabolism of proteins, although the full potential of this aspect of the method has not yet been realized. Covalent attachment of a ligand to a receptor can block the binding site to fresh ligand or, if the site is allosteric, fix a protein in an active or inactive conformation. Staros and Knowles (1978) have selectively inactivated the dipeptide transport system of *E. coli* by irradiating living cells with a photoaffinity reagent. Galardy et al. (1980) have produced irreversible activation of pancreatic secretion with a photolabile analog of cholecystokinin. More recently, persistant activation of steroidogenesis occurred when adrenocortical cells were photolysed with an arylazido-ACTH derivative (Ramachandran et al., 1981), and irreversible activation of adenylate cyclase has been produced with photoactivatable derivatives of parathyroid hormone and glucagon (Draper et al., 1982; Demoliou-Mason and Epand et al., 1982). Schaltmann and Pongs (1982) were able to follow the movement of a photoaffinity labeled steroid hormone receptor from the cytoplasm to the nucleus of the cell, and Berhanu et al. (1982) have observed the fate (internalization followed by proteolytic cleavage) of a covalent insulin–receptor complex that had been formed photochemically.

1.2.5. Time-dependent photoaffinity labeling

Park et al. (1982a,b) have examined the mechanism of promoter selection by *E. coli* RNA polymerase on T7 bacteriophage DNA, using rapid mixing and flash photolysis (which activates unmodified DNA). Different delay intervals before photolysis were used with a resolution of ~ 30 ms, limited by the mixing time. By determining the distribution of the enzyme along the DNA at different timepoints, and by showing that the enzyme exchanged relatively slowly between DNA molecules, it was demonstrated that the promoter is found by linear diffusion along the DNA, rather than by random collisions.

1.2.6. Further uses of photogenerated reagents

Extensions of the technique of photoaffinity labeling have proved useful in almost every area of biochemistry and molecular biology. In the membrane field, hydrophilic photoactivatable reagents have been used to label peripheral membrane proteins and the exposed surfaces of integral proteins while hydrophobic reagents have been used to derivatize the regions of integral proteins associated with the lipid bilayer (Chapter 6). For example, Khorana's group has developed a collection of phospholipids containing photoactivatable groups which, in membranes, are buried within the bilayer (Radhakrishnan et al., 1980; Robson et al., 1982). On activation the photolabile lipids react with neighboring lipids and proteins. Using such reagents the organization of protein complexes with respect to the lipid bilayer can be investigated.

Photochemical reagents have been devised for crosslinking both soluble and particulate proteins (Chapter 5). In a recent study, Johnson et al. (1981) were able to crosslink radiolabeled glucagon to its receptor, merely by adding a bifunctional reagent that first reacted with amino groups and could subsequently be induced to form crosslinks by photolysis.

Underivatized nucleic acids will partake in intermolecular photochemical reactions and there is an extensive literature on the use of photochemical crosslinking in the nucleic acid field. Recent work has been directed towards the development of selective methods so that the investigator can have more control over the site and nature of the crosslinks. Cantor and his coworkers are developing psoralen derivatives for protein–nucleic acid crosslinking (Cantor, 1980). Psoralens intercalate into double-stranded nucleic acids and on irradiation they form crosslinks, by addition to pyrimidine bases, in sequences where purines and pyrimidines alternate. Irradiation at the long wavelength edge of their absorption results in monoadducts rather than crosslinks. The appropriate psoralen derivative may first be attached chemically to a protein of interest and the derivative can be used for labeling nucleic acid molecules that associate with the protein.

CHAPTER 2

Photogenerated reactive intermediates and their properties

This chapter describes the reactive intermediates that are generated from photoactivatable reagents. The description is divided into three sections covering carbenes, nitrenes, and miscellaneous reactants including radicals and excited states. In keeping with the pragmatic approach of this series the chemical properties of the intermediates are described somewhat superficially. As many of the reactions described are poorly, if at all understood, this is not as unsatisfactory as it may seem. Indeed, many photogenerated reagents have proved extremely useful in the absence of any detailed knowledge of the way in which they react!

2.1. Carbenes

Carbenes, which may be formed photochemically from precursors such as diazo compounds and diazirines, are highly reactive entities containing divalent carbon. The first photogenerated reagent was a carbene, formed by irradiation of diazoacetyl chymotrypsin (Singh et al., 1962).

The typical reactions of a carbene are shown in Fig. 2.1. Clearly, such an intermediate is far more reactive than the usual electrophilic reagents used for affinity labeling and other forms of chemical modification (Means and Feeney, 1971; Glazer et al., 1975). Besides reacting with nucleophilic groups, carbenes are capable of reacting by insertion with saturated hydrocarbons and by addition to unsaturated hydrocarbons including aromatic molecules. None of the twenty different amino acid side chains are immune to attack and hence carbenes are useful not only for labeling enzyme active sites, which are expected to contain nucleophiles, but also

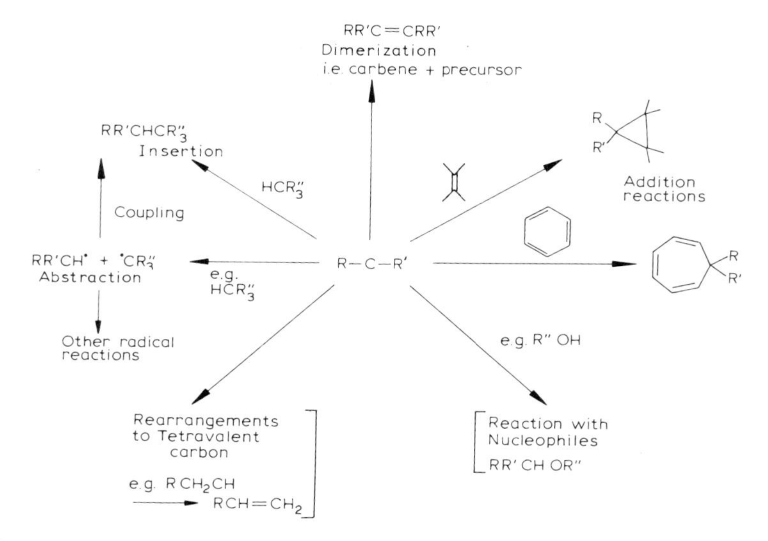

Fig. 2.1. Typical reactions of a carbene.

other receptors that may primarily contain hydrocarbon residues in their binding sites. Similarly lipids, nucleic acids, and oligosaccharides are also expected to react.

In model reactions, amide bonds will react with carbenes to form imidates (Fig. 2.2; White et al., 1978). Imidates are rapidly hydrolyzed at pH 5 but as yet there have been no reports of polypeptide chain cleavage by photoaffinity reagents. Dimerizations of the reagent (Fig. 2.1) are less important in biological photolabeling studies than during carbene formation in the gas phase or in concentrated solution in inert solvents. Rearrangements are important both as sources of unwanted long-lived reactive intermediates, and of inert byproducts. As the rearrangement products are usually peculiar to a particular class of carbene or a particular precursor they are discussed in detail in Chapter 3.

While carbenes are extraordinarily reactive it is important to remember that they exhibit an appreciable degree of selectivity in reactions with

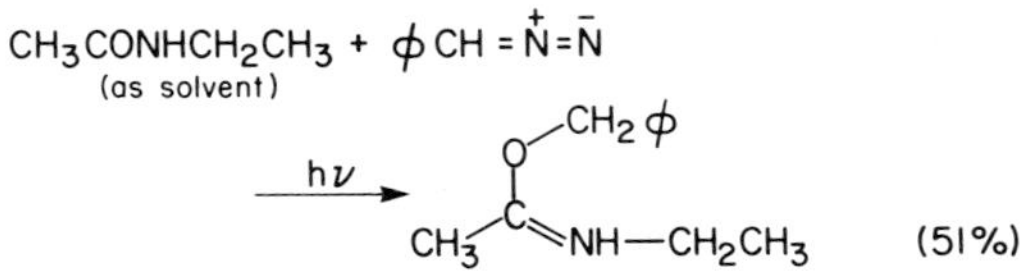

$$CH_3CONHCH_2CH_3 + \phi CH = \overset{+}{N} = \overset{-}{N} \xrightarrow{h\nu} CH_3C(OCH_2\phi){=}NH{-}CH_2CH_3 \quad (51\%)$$

(as solvent)

Fig. 2.2.

various functional groups. This selectivity depends upon the molecular structure of the carbene. Even 'the most indiscriminant species known in organic chemistry', photogenerated singlet methylene (Doering et al., 1956), reacts with nucleophiles more readily than with carbon–hydrogen bonds. In *t*-butanol, reaction with the hydroxyl group is 11 times more rapid than with a carbon–hydrogen bond, and in methanol reaction with the hydroxyl group is 22 times faster (Kerr et al., 1967). At the other end of the spectrum of reactivity, carbenes such as CF_2 are highly selective in their reactions with alkenes substituted with groups with different electronic properties, and do not insert at all into carbon–hydrogen bonds. Carbenic selectivity towards carbon–hydrogen bonds has been summarized by Atherton and Fields (1968), and selectivity towards alkenes by Moss (1980; see also the reviews listed below). Unfortunately a detailed study of the reactivity of carbenes towards the functional groups present in biological systems has not been made. The problem will receive more attention later when experiments such as hydrophobic membrane labeling that require highly indiscriminate reagents, and the problem of pseudo-photoaffinity labeling are discussed.

Research on the different properties of singlet and triplet carbenes has undergone a revival with the advent of nano- and picosecond kinetic methods. Previously the reactions of the two species have often not been carefully distinguished. Singlet carbenes are believed to be involved in rearrangements to fully bonded species, stereospecific reactions with σ and π bonds and the addition of nucleophiles, while triplets partake in abstractions, non-stereospecific addition to π bonds and other radical-like reactions. However, with the discovery of singlet–triplet equilibria and singlet carbenes, such as fluorenylidene, that undergo non-stereospecific reactions with some olefins it is apparent that we still have much to learn.

The reader who wishes to design new photogenerated reagents would be well advised to keep up with recent advances in this area (see for instance Eisenthal et al., 1980; Zupancic et al., 1981; Wong et al., 1981) but first obtain a solid background from the review literature (useful reviews are: Kirmse 1971; Moss and Jones 1973, 1975, 1978; Wentrup, 1979; Turro, 1979). In reading the literature care should be exercised in distinguishing between the reactions of photogenerated carbenes in solution or the solid state at room temperature which are relevant to us and, for instance, thermal reactions in the gas phase or photolyses in argon matrices which may follow different courses.

2.2. Nitrenes

Nitrenes (R–N) are molecules that contain monovalent nitrogen and whose reactions parallel those of carbenes (Fig. 2.1). Highly reactive species such as carboethoxynitrenes (N–COOR) which insert efficiently into carbon–hydrogen bonds and add to double bonds, in intermolecular reactions, do exist but practical considerations (see Chapter 3) have led to the almost exclusive use of aryl nitrenes which are considerably less reactive than aryl carbenes and many other classes of nitrenes. For instance, when phenyl azide is irradiated to yield phenyl nitrene in a solution of an alkane *no* insertion occurs into solvent molecules (Reiser and Leyshon, 1971). Singlet aryl nitrenes rearrange to less reactive but nevertheless strongly electrophilic species (benzazirines and cycloheptatetraenes) which, by reacting with nucleophiles on the receptor, may be responsible for much of the labeling seen with arylazido photoaffinity reagents (see Section 3.2.4). Singlets that undergo intersystem crossing to the triplet state are converted to anilines by hydrogen atom abstraction in their predominant intermolecular reaction. Such a reaction would result in a failure to couple to the receptor. However, aryl nitrenes undergo many of the reactions of Fig. 2.1 intramolecularly. For example, photochemical intramolecular insertion reactions are well documented (Fig. 2.3). It can be argued that photoaffinity labeling, in which a ligand is tightly bound to a receptor, is analogous to the intramolecular case.

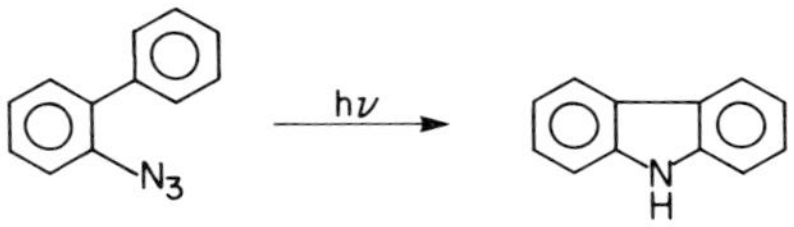

hν

N$_3$

N H

(Smith & Brown, 1951)

Fig. 2.3.

It has often been suggested that electron withdrawing groups on the aromatic nucleus may be used to increase the reactivity of aryl nitrenes. This is indeed the case in that the lifetime of a triplet nitrene in a polystyrene matrix, which is determined by the rate of abstraction of hydrogen atoms, is decreased by such substitution (Reiser and Leyshon 1970). It must be remembered though that the electrophilicity of the initially formed singlet is greatly increased by an electron withdrawing group and the substituted nitrene will react far more rapidly with nucleophiles (see for example McRobbie et al., 1976) at the expense of an alternative insertion reaction.

Despite their limited reactivity aryl nitrenes have proved exceedingly useful as photogenerated reagents and this is documented in later chapters. Reviews on nitrene chemistry include those of Lwowski (1970), Moss and Jones (1978), Wentrup (1979), Iddon et al. (1979), Colman et al. (1981), and Scriven (1983).

For an excellent summary of the potential of different classes of nitrenes as photogenerated reagents the reader should consult Lwowski (1980).

2.3. *Lifetimes of carbenes and nitrenes*

In the literature of photoaffinity labeling, much ado is made about the half-lifes of carbenes, nitrenes and other reactive intermediates. It is often implied that the half-life has a fixed value for each intermediate, but it is of course a function of the temperature and environment.

Most of the situations considered in this book approximate to pseudo first-order reactions. If k is the second-order rate constant for the reaction in solution of the activated reagent, L^*, with a functional group, F, then the $T_{1/2}$ of L^* tightly bound to a receptor site containing F will be:

$$T_{1/2} = \ln 2/F_a k$$

where F_a is the apparent concentration of F. The value of F_a cannot be easily evaluated. The concentration of hydroxyl groups in pure ethanol is 17.2 M. F_a will often be lower than this if a single nucleophile of ordinary reactivity is at the binding site. The apparent F_a can also be much larger than this if the reactivity of a nucleophile at the binding site is high because of a favorable orientation or proximity effect. We assume below that F_a is 1.0 M and the lifetimes we obtain are probably longer than those of intermediates generated on receptors.

The paucity of literature on the absolute reaction rates of carbenes and nitrenes does not permit the choice of examples particularly relevant to our problems, but some illustrative cases, calculated from data in the literature, are given here. The half-life of an aryl nitrene in 1.0 M ethanol (triplet: H atom abstraction) can be estimated to be 0.1 to 5 ms (Reiser et al., 1968; Reiser et al., 1966). Intramolecular carbazole formation from singlet 2-nitrenobiphenyl occurs with $T_{1/2} \sim 0.5$ ms (Sundberg et al., 1975), but other singlet aryl nitrenes may rearrange more rapidly (DeGraff et al., 1974), to less reactive species (Nielsen and Buchardt, 1982). In soft polystyrene matrices triplet aryl nitrenes have $T_{1/2} \sim 1$ ms; in rigid matrices lifetimes may be over 1 s (Reiser et al., 1968). Aryl nitrenes react very rapidly indeed with O_2 or with their azide precursors (Reiser et al., 1968).

Triplet diphenylcarbene has a $T_{1/2}$ of ~ 2 μs in the presence of 1.0 M isoprene (Eisenthal et al., 1980). The lifetime of singlet phenylchlorocarbene ranges from 5 to 500 ns in the presence of 1.0 M alkenes with various substituents (Turro et al., 1980). In 1.0 M methanol singlet fluorenylidene and singlet diphenylcarbene have half-lives of 0.77 ns and 0.02 ns respectively (Zupancic and Schuster, 1980; Eisenthal et al., 1980). Triplet carbenes react very rapidly with 0_2.

These data at least give us a feeling for the life-times of photogenerated carbenes and nitrenes, and confirm the idea that carbenes are more reactive than related nitrenes. The life-times of less-commonly used reactive intermediates may also be approximated using pseudo first-order kinetics, unless reaction occurs from an excited state, when relaxation to the ground state will often be limiting.

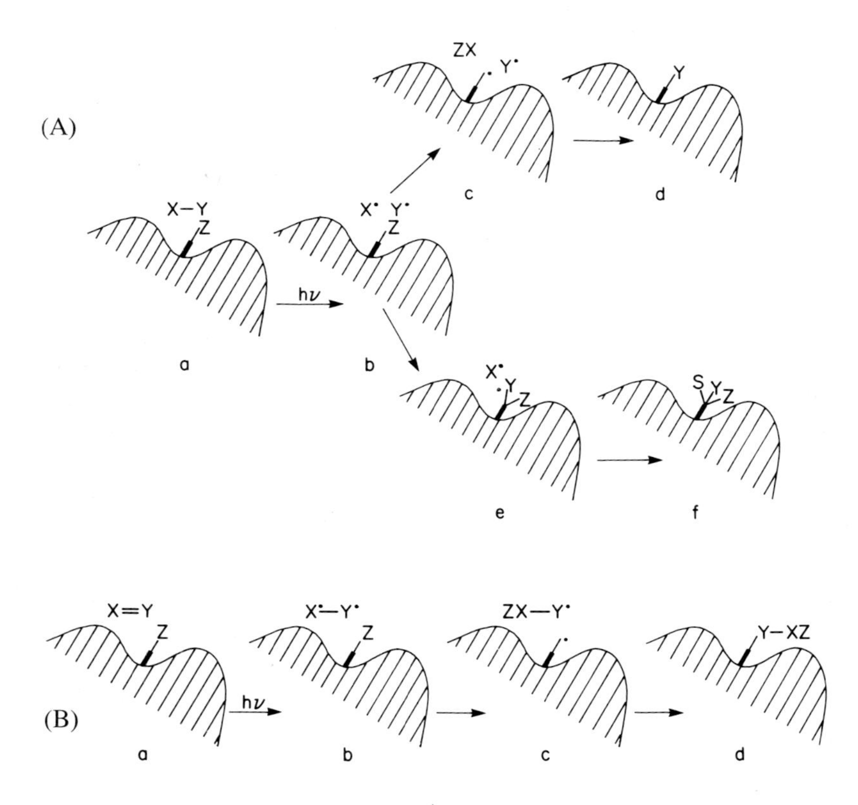

Fig. 2.4. Reactions of radicals generated from a ligand bound to a receptor. X–Y or X=Y is the photoactivatable ligand; Y is radiolabeled but X is not; Z is an atom in the ligand binding site (e.g. H); S represents termination of the reaction by H˙, by another radical from the buffer, or by X˙. Elimination of a radical from the receptor could also terminate the reaction. If termination occurred at stage *c*, a covalent adduct with a radiolabeled fragment of XY would not be formed but the receptor might be inactivated (see e.g. Ogez et al., 1977; Smith and Benisek, 1980).

2.4. *Free radicals and excited states*

Specific examples of the classes of carbenes and nitrenes useful for photoaffinity labeling are given in Chapter 3. Covalent bonds to biological molecules have also been formed in reactions with free radicals and excited states. As these species are used more rarely their chemistry is outlined in this chapter only.

Free radicals, formed by homolytic bond cleavage, may couple to groups on a receptor in several ways (Fig. 2.4A). Diradicals (e.g. triplet excited states) probably abstract an atom from the receptor (e.g. $H^{\cdot}$) and the two radicals that are formed couple (Fig. 2.4b).

2.4.1. α,β-Unsaturated ketones

Galardy et al. (1973) introduced acetophenones and benzophenones for photoaffinity labeling (see also Martyr and Benisek, 1973; Katzenellenbogen et al., 1974). The properties of excited states generated from these and other α,β-unsaturated ketones are well understood (Turro, 1979). In most cases a triplet excited state is formed that abstracts a hydrogen atom from a donor yielding two radicals which subsequently couple. Unreacted excited species relax to the ground state and may be excited repeatedly until they react (Fig. 2.5).

The bond strength of the donor is crucial in determining its reactivity. In general, the weaker the bond to the donor the more readily the $H^{\cdot}$ atom is transferred to the excited ketone. Galardy et al. (1973) were able to demonstrate that reaction with the relatively weak α-CH bonds of amino acids would occur in the presence of water, a useful property for a labeling reagent in a site exposed to or containing solvent.

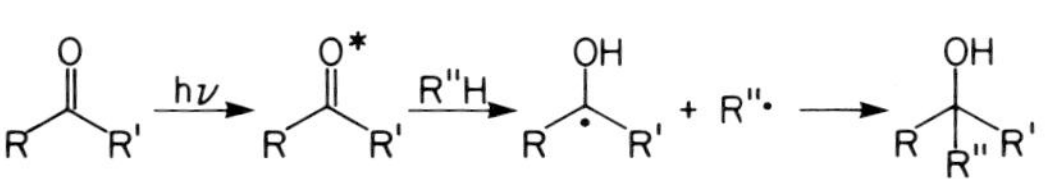

Fig. 2.5.

TABLE 2.1
Some less commonly used reagents

Class	Example	Receptor	Comments/References
α,β-Unsaturated ketone	Ecdysterone	Labeling of insect chromosomes	Simple α,β-unsaturated ketone. Visualized at labeling site by immunofluorescence (Gronemeyer and Pongs, 1980).
α,β-Unsaturated ketone	17β-Hydroxy-4,6-androstadien-3-one	Androgen binding protein	Dienone (λ_{max} 345 nm, ε300,nπ*). Photoinactivation and photolabeling of receptor (Taylor et al., 1980).
α,β-Unsaturated lactone	Cardiotonic Steroids	Na,K-ATPase	Labeling not via lactone but to excited Trp or Tyr residues on protein? Poor efficiency: extent of labeling limited to 1% by photocrosslinking of enzyme subunits. (Forbush and Hoffman, 1979).
Arylhalide	5′-iodo-2′-deoxyuridine monophosphate	Thymidine kinase	Both iodine and nucleotide covalently incorporated. Only nucleotide incorporation parallels inhibition of enzyme activity (Chen et al., 1976).
Nitroaryl compounds	Flunitrazepam	Benzodiazepine	Labels 50,000 MW polypeptide. Also autoradiographic localization in tissue slices (Mohler et al., 1980).
Purine	ATP	Ile tRNA synthetase	Irradiation at 254 nm. Labeling yield limited to 15% by photoinactivation of the enzyme (Yue and Schimmel, 1977).

Pyrimidine	pUp	Ribonuclease	Acetone sensitized photolysis. Labeled residues identified (Havron and Sperling, 1977).
Unusual nucleic acid base	Poly-4-thiouridylic acid	*E. coli* ribosomes	Crosslinking to protein 1 of 30 S subunit. Requires O_2 for attachment (Fiser et al., 1974).
Photochemical nucleophilic aromatic substitution	4-Nitrophenyl-α-D-galactopyranoside	Lactose transporter of *E. coli*	Non-specific labeling prevented by anaerobic conditions. Azidonitro phenyl analog gave much non-specific labeling (Kaczorowski et al., 1980).
N-Nitroso compound	methyl(acetoxymethyl)-nitrosamine	Acetylcholine esterase	Selective destruction at active site because of proximity to acidic residue which protonates $-N{=}O$ and enhances photosensitivity. Not generally useful (Eid et al., 1981).
Diazonium salt	*p*-Dimethylaminobenzene diazonium salt	Acetylcholine esterase	Use limited by chemical reactivity of ArN_2^+. Reaction via $Ar\cdot$ and Ar^+? (Goeldner and Hirth, 1980).
Triarylethylene	Estrogen analog	Estrogen receptor	Up to 50% photochemical attachment (Katzenellenbogen et al., 1974).
Photo-oxidation	Dyes/O_2	E.g. serum albumin	Low efficiency, high dye concentration (Brandt et al., 1974).

While it has been assumed (Galardy et al., 1973) that the reactive species is the $n\pi^*$ excited state this may not always be the case. For acetophenones the nature of the photogenerated triplet is highly solvent dependent (Turro, 1979). In more polar environments the species formed is the $\pi\pi^*$ state which is less reactive than the $n\pi^*$ state and in certain cases it may prove useful to substitute the molecule with an electron withdrawing group (e.g. CF_3-) on the aromatic ring to ensure formation of the desired T_1 ($n\pi^*$) intermediate.

A detailed study of the relative merits of acetophenone, benzophenone and azidoaryl derivatives was made by Galardy et al. (1974). Sensitive amino acids such as Cys, Met, Tyr and Trp were destroyed on irradiation in the presence of the ketones but not the azides. Nevertheless, benzophenone derivatives which can be irradiated at $>$ 320 nm have been occasionally used with success (for a recent application see Williams and Coleman, 1982).

Further examples of α,β-unsaturated ketones include molecules such as steroid derivatives that were not originally designed as photoaffinity reagents. Polyunsaturated compounds that may be photolysed efficiently at wavelengths above 300 nm are preferred (e.g. Fig. 2.6: Dure et al., 1980; Nordeen et al., 1981; Sadler and Maller, 1982).

While true photoaffinity labeling has been achieved with α,β-unsaturated ketones (Table 2.1, and above), it should be noted that Benisek and coworkers have observed irreversible inhibition of ketosteroid isomerase from two sources without covalent attachment of the reagent (Ogez et al., 1977; Smith and Benisek, 1980). In each case stoichiometric modification of a residue at the active sites was detected (see Fig. 2.4, Legend).

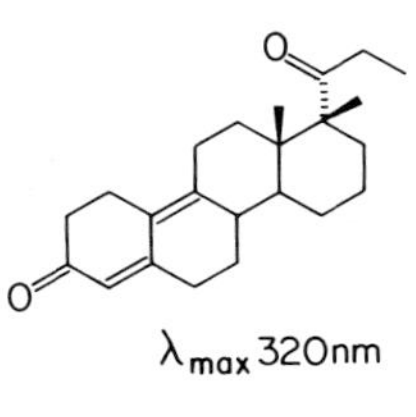

λ_{max} 320nm

Fig. 2.6.

2.4.2. Aryl halides

Aryl halides have been used with moderate success as photoaffinity reagents and they react in a process initiated by homolytic fission at the carbon–halogen bond (Sharma and Kharash, 1968, Grimshaw and de Silva, 1981).

Examples include bromo- and iodouridine derivatives (Fig. 2.7: Lin and Riggs, 1974; Chen et al., 1976) and iodohexestrols (Katzenellenbogen and Hsiung, 1975). The latter were not useful for derivatizing the estrogen receptors; neither ^{125}I nor ^{3}H from labeled derivatives were incorporated into protein. The photochemical incorporation of ^{14}C from 5-iodo-2′-deoxyuridine monophosphate into thymidine kinase did parallel inactivation of the enzyme. In this case some ^{125}I was also incorporated but the extent of incorporation did not parallel inactivation (Chen et al., 1976). Recently, radiolabeled thyroxine and 3,5,3′-triiodothyronine have been used to label thyroid hormone receptors (Van der Walt et al., 1982). In this case, the iodine released did not react with protein.

2.4.3. Nitroaryl compounds

Nitroaryl compounds including amino acid and peptide derivatives (Escher and Schwyzer, 1974), chloramphenicol (Sonenberg et al., 1974: which gave much non-specific labeling), and flunitrazepam (Mohler et al., 1980) have been used to photolabel receptors and they probably react via triplet

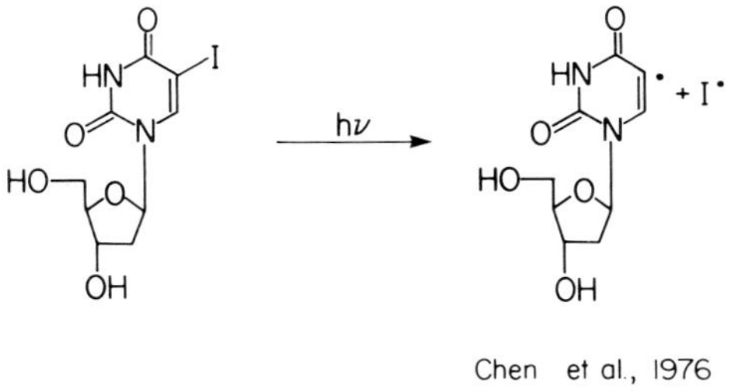

Chen et al., 1976

Fig. 2.7.

excited states with diradical like properties (Turro, 1979). Nitro groups present in otherwise photoactivatable molecules including those used for photochemical nucleophilic aromatic substitution (see Section 2.4.5) should not be ruled out as the origin of covalent attachment.

2.4.4. Purines and pyrimidines

The photochemistry of purines and pyrimidines has been extensively investigated. It is a field in itself, beyond the scope of this book. A variety of reactions that may occur directly from singlet or triplet excited states or with free radical intermediates can result in inter- or intrastrand nucleic acid crosslinking or the attachment of nucleotide bases to proteins (for reviews on nucleic acid photochemistry, see Smith, 1976; Wang, 1976; Jakoby and Wilchek, 1977; as well as recent issues of *Photochemistry and Photobiology,* and *Photochemical and Photobiological Reviews.* The scholarly article of Shetlar (1980) reviews light-induced protein–nucleic acid crosslinking).

These reactions usually proceed, at short wavelengths, with relatively low quantum yields and the efficiency may be improved by taking advantage of synthetic or naturally occurring photosensitive bases, which often absorb at longer wavelengths. Examples are: 4-thiouridine (Fiser, et al., 1974), wybutine (Matzke, et al., 1980), 5-carboxymethoxyuridine (Ofengand et al., 1982), and azidopurines (Czarnecki et al., 1979; Cartwright and Hutchinson, 1980; MacFarlane et al., 1982).

2.4.5. Photochemical aromatic substitution

Nitrophenyl ethers capable of undergoing photochemical nucleophilic aromatic substitution (Cornelisse et al., 1975, 1979; Cornelisse and Havinga, 1975; Havinga and Cornelisse, 1979) are recent additions to the list of photochemical reagents (Jelenc et al., 1978: Fig. 2.8).

Reaction occurs via a shortlived triplet excited state, at rather low quantum efficiency, and appears to be quite selective for amino groups (Jelenc et al., 1978). This reaction has been used to label a galactose

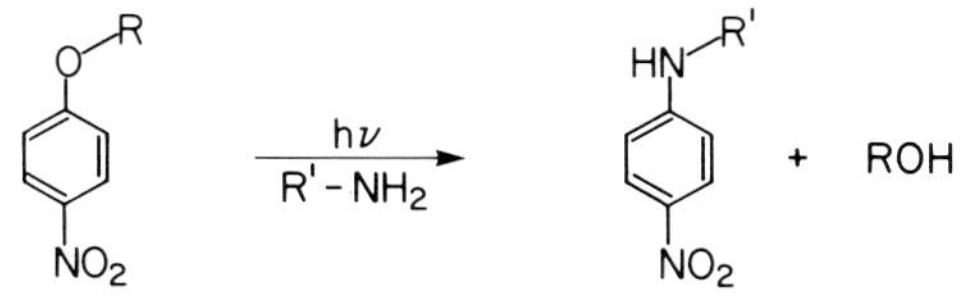

Fig. 2.8.

transport system with 4-nitrophenyl-σ-D-galactopyranoside (Kaczorowski et al., 1980).

2.4.6. *Psoralens*

The last major group of reagents are the psoralens which have earlier been exploited to form crosslinks in double-stranded DNA and RNA (Song and Tapley, 1979; Parsons, 1980; Hearst, 1981). Reaction occurs primarily by 2 + 2 addition to pyrimidines but also with purines (Fig. 2.9).

The demonstration of stepwise adduct formation by psoralens was an important advance. At 390 nm monoadduct formation occurs; a crosslink is completed by irradiation at 360 nm (Chatterjee and Cantor, 1978). DNA-psoralen monoadducts have been used to covalently label complementary sites on large RNA molecules. The sites may then be mapped by electron microscopy of the denatured nucleic acid complexes (Wollenzien and Cantor, 1982). Psoralen derivatives for attachment to proteins have been made (Cantor, 1980) and this will allow the exploration of protein binding sites on nucleic acids by photochemical crosslinking.

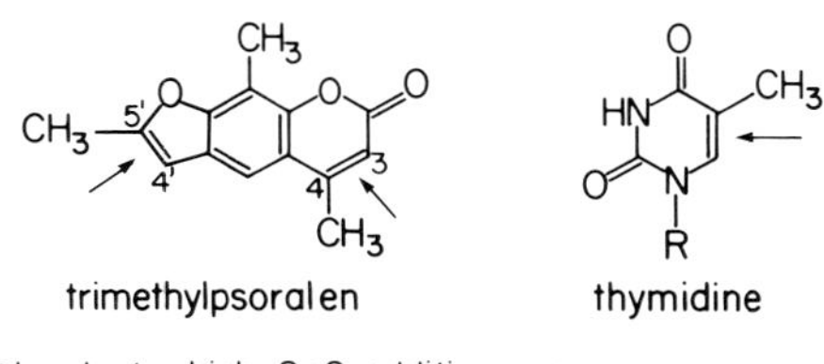

Fig. 2.9.

2.4.7. *Miscellaneous*

Miscellaneous examples of photogenerated reagents include triarylethylenes (Katzenellenbogen et al., 1974), *N*-nitroso compounds (Eid et al., 1981) and diazonium salts (Goeldner and Hirth, 1980). Several naturally occurring ligands such as pyridoxal phosphate (Greenwall et al., 1973), bilirubin (Hutchinson and Mutopo, 1979), and α-amanitin (Lutter, 1982) have also been used. As the major emphasis in this monograph is on the more conventional carbene and nitrene precursors which will be described in detail in Chapter 3, a compilation of examples of the less usual reagents is given in Table 2.1, and there will be little further discussion of their properties. When such molecules are readily available, in labeled form, perhaps as natural products or drugs, the opportunity should be taken to test them as photoactivatable reagents. But, if carbene or nitrene precursors are also conveniently obtained they would usually be the reagents of choice as they may be photolysed at suitable wavelengths with high efficiency (Chapter 3). While a few of the less common reagents may be photolysed efficiently this is not generally true and in some cases the conditions used are so harsh that it is not clear whether the primary photochemical event is on the receptor or the ligand. For instance, while several useful diazo and azido derivatives of cardiotonic steroid inhibitors of Na,K-ATPase have been made, it is possible to perform photoaffinity labeling at low efficiency with naturally occurring ligands (Forbush and Hoffmann, 1979). The photoreactive group might be supposed to be the α,β-unsaturated-δ-lactone group (Fig. 2.10), but the action spectrum for labeling suggests that the protein is the photoactivated species.

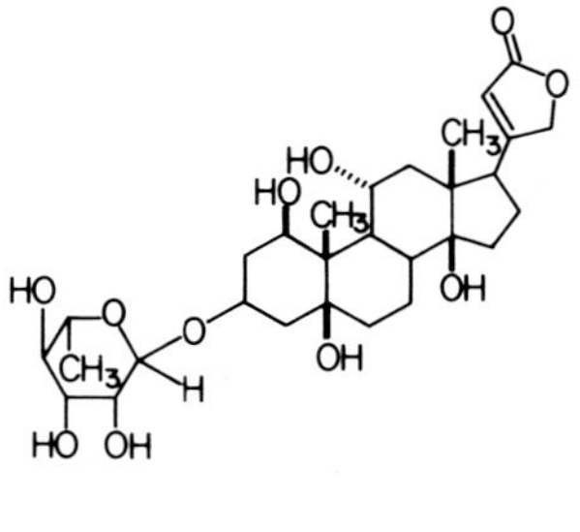

Fig. 2.10.

In general, it is advisable to perform photochemical labeling in the absence of oxygen (Chapter 4). Occasionally, with the less common reagents, a requirement for the presence of oxygen has been noted, e.g. for the reaction of 4-thio-UMP with ribonuclease (Sawada and Kanbayashi, 1973), and when covalent attachment of dyes occurs during photo-oxidation experiments (Brandt et al., 1974).

2.5. *Photosuicide inhibitors*

Goeldner, Hirth and colleagues have presented evidence that the efficiency of labeling by certain photoaffinity reagents is improved in the environment provided by the receptor. They define a photosuicide inhibitor as a 'ligand analog of an enzyme or a receptor, the photodecomposition of which is selectively induced by the intrinsic physico-chemical properties of an active site' (Goeldner et al., 1982).

Two forms of photosuicide inhibition have been described. In the first, the reagent is activated by energy-transfer from a tryptophan close to the ligand binding site. For example, acetylcholine esterase was labeled with the photoactivatable ligand *p-N,N*-dimethylaminobenzene diazonium fluoroborate. The labeling rate was increased under conditions where energy transfer occurred (Goeldner et al., 1980). In the second form of photosuicide inhibition the properties of the active site directly affect the photosensitivity of the reagent. The sensitivity of *N*-nitrosoamines to light of 330 to 340 nm is greatly increased in weakly acidic media with low dielectric constants. Analogs of acetylcholine containing the *N*-nitroso group have been used to label acetylcholine esterase and the action spectrum for labeling suggests that the analogs are protonated in the active site (Eid et al., 1981).

By using active site residues to promote the photolysis of a ligand, it should be possible to increase the specificity of photoaffinity reagents, allowing the use of weakly or loosely bound reagents that would usually give much non-specific labeling. Unfortunatly, the analogy between chemical and photochemical suicide reagents is tenuous. In the former, the reactive species is generated by catalytic turnover allowing the prediction of suitable reagents based on our knowledge of enzyme chemistry. Photo-suicide inhibition as defined above has the limitation that it is not possible to predict the amino acid side chains in the vicinity of most receptor binding sites. Further, non-specific labeling might also be promoted by the mechanisms given above.

CHAPTER 3

Reagents for photoaffinity labeling

In Chapter 2 an account was given of the various classes of molecules that have been used as photoactivatable reagents. This chapter concentrates on the three classes, aryl azides, diazo compounds and diazirines, which can be considered as proven and established.

3.1. Requirements for a useful photoactivatable reagent

An ideal photoaffinity labeling reagent would meet the criteria listed below, and subsequently discussed in more detail for each class of reagent.

(i) The reagent should be stable not only to storage but to the conditions under which the experiments are performed. Of particular importance in biochemical experiments are the prevailing pH and the reducing potential of the environment.

(ii) The reagent should be simple to synthesize. Unless it is formed or added at the last step, the photolabile group may be required to withstand harsh treatments such as oxidation and reduction, extremes of pH and exposure to powerful nucleophiles and electrophiles. A further considera-tion is that the reagent will likely be needed in radiolabeled form at high specific radioactivity.

(iii) The modified ligand should closely resemble the original molecule. Gross alterations to a ligand such as the addition of a bulky photolabile group can unacceptably reduce its affinity for a receptor.

(iv) The photolabile group should possess appropriate photolysis pro-perties. First, it should be activated at wavelengths that do little or no damage to the other components of the system. In general, this means that

TABLE 3.1
Some nitrene and carbene precursors that have been used to label biological molecules

Reagent	Structure	ε	Long-lived intermediates	Comments/references
1. *Nitrene precursors azides*				
Aryl azides	N_3, R	$>10^4$ (250 nm) with shoulders at 277 and 286 nm for R = alkyl. 2-nitro-4-azido anilines have ε ∽ 5,000 (460 nm) but ψ is low in the visible region. See also Section 3.2.3.	For example N (see text)	The most popular of photoaffinity reagents. See Fleet et al. (1969) and the many references in Bayley and Knowles (1977) and this work. Many bifunctional derivatives are listed in Table 5.1.
8-Azidopurines	NH_2, N, N, N, N, N_3, R	13,000 (280 nm)	None known	Adenine, guanine and inosine derivatives have been made. Used successfully to label many nucleotide binding sites but derivatives do not bind to all nucleotide receptors. See Casnellie et al. (1978); Geahlen et al. (1979); Czarnecki et al. (1979).
Alkoxycarbonyl-azides	$ROCON_3$	Low 90 (250 nm) 2.5 (300 nm)	None known	Highly reactive nitrenes (Lwowski, 1970). Rarely used because of low ε, but see Vaver et al. (1979).
Alkylazides	RN_3	25 (290 nm)	Imine formation predominates (but see Stoffel et al., 1982; Stoffel and Metz, 1982)	Lwowski (1970). Have been incorporated into phospholipids (Stoffel et al., 1978). Not generally recommended because of rearrangements and low ε.

Table 3.1 continued

Reagent	Structure	ε	Long-lived intermediates	Comments/references
Phosphorylazides	$(RO)_2P(=O){-}N_3$	Low 20 (250 nm) 316 (208 nm)	None known	Highly reactive nitrenes. See Breslow et al. (1974).
Azidophosphates	$RO{-}P(=O)(O^-){-}N_3$	Probably low in near UV	None described	E.g. α-azido AMP and α-azido GMP. A detailed study of the photochemistry of these compounds has not yet been made. See Chládek et al. (1977).
Arylsulfonyl-azides	$R{-}C_6H_4{-}S(=O)_2{-}N_3$	R = alkyl absorb below 300 nm.	None described	Pyrene sulfonylazide (λ_{max} 355 nm) has been used as a hydrophobic reagent. Sator et al. (1979).
2. *Carbene precursors*				
(a) *Diazo compounds*				
Diazocarbonyl derivatives	RCN_2COR_1	∽ 10^4 (240-280 nm) but <50 at ∽ 350 nm	$RR_1C{=}C{=}O$ Wolff rearrangement	Extent of rearrangement depends on RR_1 (see Chowdhry and Westheimer, 1979). Now seldom used (for a list of examples see Bayley and Knowles, 1977).
Trifluoromethyl-diazoacetates	CF_3CN_2COR	Again low at ∽ 350 nm	Wolff rearrangement	Improved stability. Extent of rearrangement greatly reduced. Synthesis and properties, Chowdhry et al. (1976); Use in a phospholipid analog; Robson et al. (1982).

Table 3.1 continued

Reagent	Structure	ε	Long-lived intermediates	Comments/references
p-Toluene-sulfonyldiazo-acetates	$CH_3\psi SO_2CN_2COR$	140 (340 nm)	Wolff rearrangement	Chowdhry and Westheimer, (1978). Improved ε in near UV and reduced extent of rearrangement. Rather bulky.
(Dansyldiazo-methyl)phosphinates	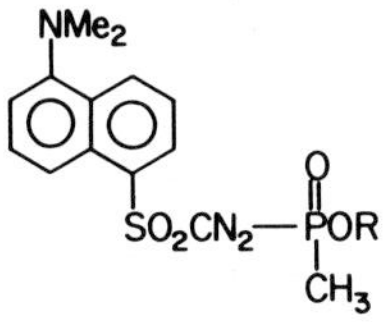	4×10^3 (350 nm)	Wolff rearrangement	Stackhouse and Westheimer (1981). Greatly improved ε. Fluorescent, bulky.
(b) *Diazirines*				
3-H-3-aryldiazirines	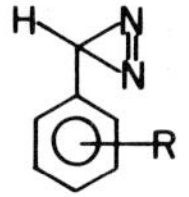	200-300 (350-400 nm)	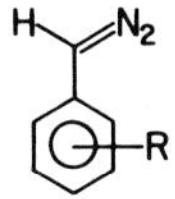	Difficult synthesis: Smith and Knowles (1973, 1975). Applications: Bayley and Knowles (1978b); Robson et al. (1982); Huang et al. (1982).
Trifluoromethyl-phenyldiazirines	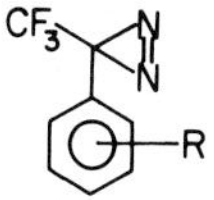	266 (353 nm)	None	The diazo rearrangement product is inert. A derivative for reaction with nucleophiles has been made. See Brunner et al. (1980). For applications see Brunner and Richards (1980); Brunner and Semenza (1981).
Adamantane diazirines	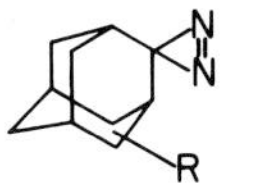	245 (372 nm)	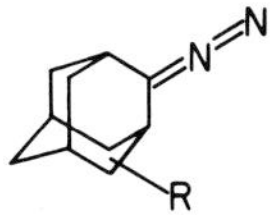	Bayley and Knowles (1978b, 1980). As yet no functionalized reagents available for use in chemical synthesis or for attachment to macromolecules.

irradiation at wavelengths absorbed by proteins and nucleic acids (< 300 nm) should be avoided. Second, photolysis should be highly efficient; that is the reagent should have a high extinction coefficient (ε) and the products should be formed in high quantum yield (ψ). If ε and ψ are large, photolysis at wavelengths < 300 nm can often be tolerated.

(v) The photogenerated reactive intermediates should be highly reactive (short-lived). Carbenes and nitrenes generated by photolysis are usually reactive enough for most applications. A major problem is to avoid less reactive intermediates formed by the rearrangement of more reactive species or their precursors.

(vi) The reactive intermediates should form stable adducts with the receptor. When a label must be located in a small peptide after degradation of the receptor protein the adduct must withstand rough handling, particularly if chemical cleavage is used. When a labeled polypeptide is merely identified, e.g. by polyacrylamide gel electrophoresis, the requirements for stability will be less stringent.

A list of useful reagents is given in Table 3.1. Few if any fulfill all the requirements listed here but several come close to doing so.

3.2. Properties of aryl azides

Aryl azides which were introduced as photoaffinity reagents by Fleet et al. (1969) are now the most commonly used photoactivatable reagents. By early 1976 there were almost one hundred examples of their use (Bayley and Knowles, 1977) and today it would be impracticable to list a complete bibliography. The ease of synthesis of aryl azides, and not their other desirable properties, is probably responsible for their great popularity.

3.2.1. Stability of aryl azides

Many molecules with a high nitrogen content are potentially explosive. Low molecular weight aryl azides are generally stable but as undiluted solids or liquids they should be handled with respect, especially if they are prepared in gram amounts. For example, distillations should be done under vacuum at the lowest possible temperatures, and solids are best removed from containers with soft plastic spatulas.

Quite recently it was found that azides are rapidly converted to amines by the thiols that are commonly used in buffers to provide a reducing environment (Cartwright et al., 1976; Staros et al., 1978; Bayley et al., 1978). The rate of reduction is approximately one thousand times faster with a dithiol such as dithiothreitol than with a monothiol such as glutathione or β-mercaptoethanol, and with dithiothreitol a bell-shaped dependency on the pH is followed suggesting that the monoanion is the reducing agent. While the rate is maximal at a pH value of approximately 10, it is appreciable at lower pH values, for instance *N*-(2-nitro-4-azidophenyl)ethylenediamine (2 mM) has a half-life of 5 to 10 min in 10 mM DTT at pH 8.0 (30°C). Aryl azides containing electron withdrawing substituents are particularly susceptible to reduction.

It was concluded that dithiols should not be used in photoaffinity labeling experiments with aryl azides and monothiols should be used with caution. The reduction reaction can often be monitored spectrophotometrically, e.g. 2-nitro-4-azidoanilines which are red-orange (λ_{max} 480 nm in H_2O) are converted to purple diamines (λ_{max} 550 nm). TLC and IR spectroscopy are also useful (Staros et al., 1978).

Obviously the investigator should also be aware of any ground state reactions peculiar to the system under study. For example 2-azidopyridines isomerize to tetrazoles which are usually less sensitive than the azide to irradiation (Fig. 3.1). The position of the equilibrium is both temperature, solvent and pH dependent. Even if the equilibrium constant for such an isomerization in solution is known the situation in a ligand binding site cannot be predicted. Azide-tetrazole isomerization was an important consideration in experiments using 2-azidoadenine derivatives which have recently been described by MacFarlane et al. (1982).

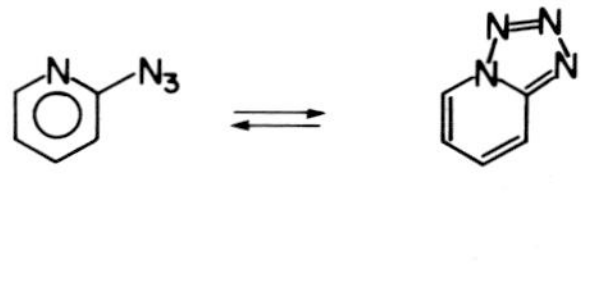

Fig. 3.1.

3.2.2. *Size of the photolabile group*

Aryl azides are rather bulky groups to append to a ligand, particularly to a small organic molecule, but it has proved possible in many cases to build the reactive group into the structure of a molecule, for instance: 8-azido-cAMP and 4-azido-phenylalanine (Fig. 3.2).

In other cases regions of molecules that are insensitive to structural perturbation have been derivatized (see below).

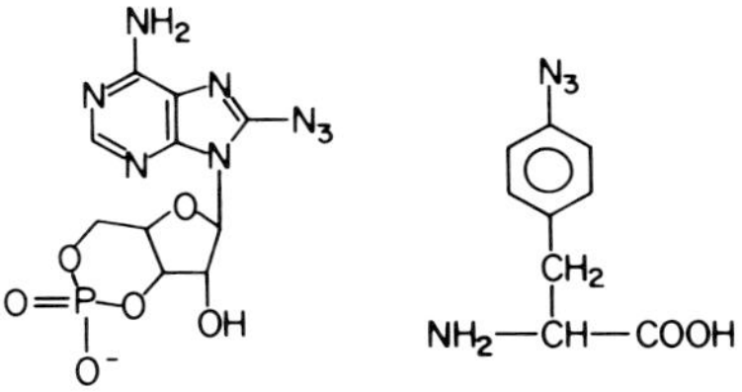

Fig. 3.2.

3.2.3. *Photolysis properties of aryl azides*

Phenyl azide itself absorbs at 250 nm ($\varepsilon \sim 10{,}000$) with the characteristic shoulders of an aryl azide at 277 and 286 nm. Electron withdrawing substituents produce a shift of the λ_{max} towards the visible, e.g. esters of *p*-azidobenzoic acid absorb at ~ 280 nm ($\varepsilon \sim 20{,}000$). Nitro substitution leads to the appearance of an absorption band at still longer wavelengths, e.g. 2-nitro-4-azidophenylalanine (λ_{max} 333 nm; ε 1,700). Further examples are given in Patai (1971).

The first azidoaryl reagent, 1-fluoro-2-nitro-4-azidobenzene (Fleet et al., 1969), was astutely based upon Sanger's reagent, 1-fluoro-2,4-dinitrobenzene, for reaction with protein amino groups. It was shown that the anilines formed from the azido reagent strongly absorb visible radiation (λ_{max} 460 nm; ε 5,000), and it was envisioned (Section 3.1 (iv)) that this would be a considerable advantage of this class of molecules. It has been demonstrated, however, that the quantum yield of photolysis of 2-nitro-4-azidoanilines is far greater in the ultraviolet than in the visible wavelength region (Matheson et al., 1977) and photolysis is usually so rapid in the ultraviolet that photochemical attachment of the ligand is often complete before appreciable damage to the receptor occurs. Indeed, simple aryl azides which absorb below 300 nm have been used with success in many photoaffinity labeling experiements, and it appears that substitution with a nitro group is not necessary except in systems unusually sensitive to ultraviolet radiation.

8-Azidoadenine derivatives (e.g. Czarnecki et al., 1979) have proved exceptionally useful and they absorb maximally at ~ 280 nm. These molecules bear a structural resemblance to carbonylazides, which yield extremely reactive nitrenes.

3.2.4. Undesirable side reactions of aryl azides

The reaction products from aryl azides in photoaffinity labeling experiments have not been fully characterized. Indeed the reactions of aryl nitrenes in solution remain a topic of current research.

In Chapter 2, it was indicated that aryl nitrenes, when compared with arylcarbenes, are relatively unreactive. Indeed, it has been suggested that they are unsuitable for labeling binding sites composed entirely of hydrocarbon residues (Bayley and Knowles, 1978a,b), although the general utility of these reagents is unquestionable. This lack of reactivity may in part be due to the intrinsic properties of triplet aryl nitrenes which have been reported to be stable when matrix isolated at temperatures as high as 77 K (Smolinsky et al., 1962), and have lifetimes (microseconds to milliseconds) in reactive solvents or soft polystyrene matrices at room temperature which are far longer than predicted for destruction by diffusion controlled events (Reiser et al., 1968; Lehman and Berry, 1973; and see Section 2.3)*.

* The major intermolecular reaction of triplet aryl nitrenes in solution is hydrogen atom abstaction to form primary amines. For a photoaffinity reagent bound to a receptor, this would result in a failure to couple. However, it is possible that the intramolecular photochemistry of aryl azides is more relevant, and here numerous examples of insertion by triplets have been noted. Presumably, these are two step processes: hydrogen atom abstraction, followed by radical coupling (cf. Figs. 2.1 and 2.3).

A second cause of the relatively low reactivity of aryl nitrenes is the rearrangement of the singlet states to azacycloheptatetraenes or benzazirines (Fig. 3.3: Chapman, 1979; Iddon et al., 1979; Colman et al., 1981;

Fig. 3.3.

Takeuchi and Koyama, 1981; Nielsen and Buchardt, 1982). These species act as electrophiles, and they (or in the case of substituted aryl azides, further rearrangement products) may have appreciable lifetimes at room temperature. For example, the intermediate generated by photolysing a *p*-alkylaryl azide had a half-life of 0.1 ms in the presence of 1 mM diethylamine (i.e. $k = 7 \times 10^6\ M^{-1}\ s^{-1}$). Recent evidence for very long-lived intermediates formed from aryl azides during photolabeling experiments include a detailed study of the labeling of arginine kinase and creatine kinase with ATP-γ-azidoanilide. When the reagent was irradiated for 6 min and *then* mixed with a kinase, inhibition occurred, in the dark with $T_{1/2}$ for onset of 3 h (Vandest et al., 1980). For further examples of long-lived species generated from aryl azides see Nielsen et al. (1978), Mas et al. (1980), and Nicolson et al. (1982); for further discussion see Staros (1980) and Section 4.7.4.

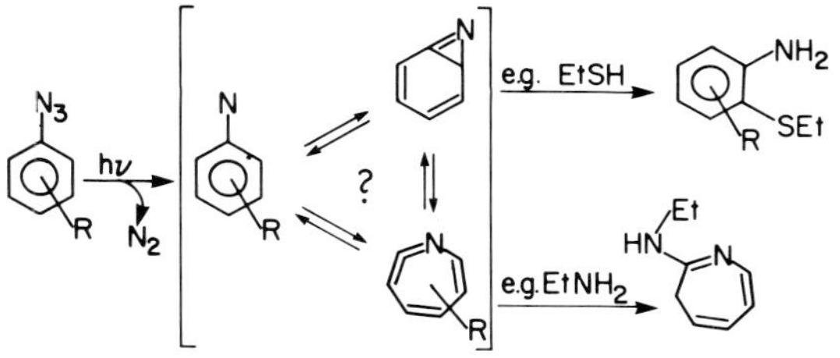

Undesirable photochemical reactions occur with many aryl azides that have ortho substituents, which should be avoided. Intramolecular reaction of the nitrene results in wastage of the photogenerated intermediates, e.g. Fig. 3.4.

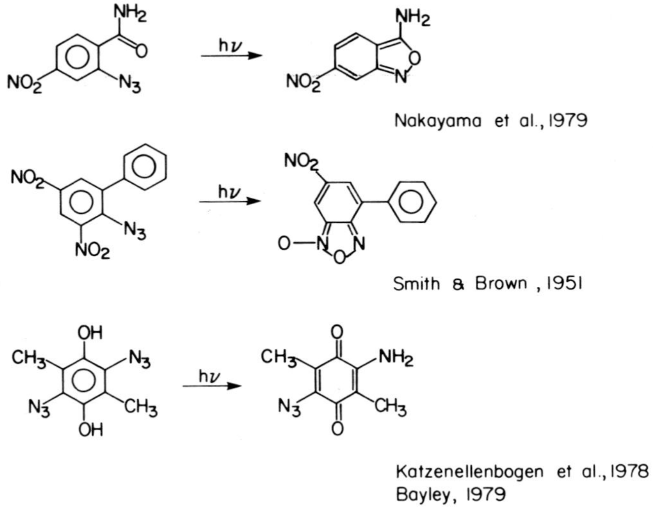

Nakayama et al., 1979

Smith & Brown, 1951

Katzenellenbogen et al., 1978
Bayley, 1979

Fig. 3.4.

3.2.5. Stability of the photoinduced crosslink

The stability of the photochemically induced linkage between a receptor and an arylazido ligand is a question that will be addressed more thoroughly in the next chapter. Based on the known photochemistry of arylazides several more or less labile products from reactions with polypeptide side-chains can be proposed. They range from carboxyazepines (from glutamic or aspartic acid) which are exceedingly sensitive to hydrolysis (Coleman et al., 1981), to 2-aminophenylsulfides (from cysteine, e.g. Fig. 3.3) which are stable molecules. In many actual cases the linkages formed have been strong enough to withstand the rigors of protein chemistry, in others a fraction of the bonds have proved to be labile.

3.2.6. Other classes of azides

Finally, it is worth considering whether other classes of azides are suitable for photoaffinity labeling (see also Lwowski, 1980). Certainly, others yield nitrenes that are highly reactive (Table 3.1). Candidates for photoactivatable reagents include alkyl azides, acyl azides ($RCON_3$), alkoxycarbonyl azides ($ROCON_3$), sulfonyl azides (RSO_2N_3) and phosphoryl azides ($(RO)_2 PON_3$).

The nitrenes derived from alkyl azides, which absorb only weakly in the ultraviolet, generally rearrange to imines which hydrolyse to aldehydes or ketones in aqueous media (Fig. 3.5). The reaction may occur directly through an excited singlet state of the azide (Kyba and Abramovitch, 1980). Very recently, Stoffel and colleagues (Stoffel et al., 1982; Stoffel and Metz, 1982) have presented evidence that alkylazido fatty acids in membranes insert into neighboring fatty acids on irradiation. Perhaps in the ordered environment of the lipid bilayer there is a conformational constraint on hydrogen or alkyl group migration permitting the formation of a reactive triplet nitrene. A similar situation could occur in the ligand binding site of a receptor.

$$RCH_2N_3 \xrightarrow{h\nu} R{-}CH{=}NH$$

Fig. 3.5.

Acyl nitrenes are highly reactive but unfortunately their precursors, acyl azides, are chemically reactive, e.g. they acylate amino groups. While sulfonyl- and alkoxycarbonyl azides are somewhat less reactive than acylazides, and the nitrenes derived from them are highly reactive (Lwowski, 1970; and for sulfonyl nitrenes see Abramovitch et al., 1981 and references therein), the azides must be irradiated in the UV, below 300 nm. Phosphorylnitrenes are the most reactive nitrenes known and will insert into CH bonds in the presence of hydroxyl groups (Breslow et al., 1974). Again, photolysis occurs at ultraviolet wavelengths.

Examples of the use of nitrenes other than aryl nitrenes as photogenerated reagents, which the reader should evaluate for himself, include alkylazido phospholipids (e.g. Stoffel et al., 1978), an arylsulfonylazide as a hydrophobic reagent for membranes (Sator et al., 1979), alkoxyacyl azido phospholipids (Vaver et al., 1979), and γ-azidophosphoryl nucleotides (Chladek et al., 1977). Of course, the latter ($ROPO_2N_3^-$) may well have a completely different photochemistry to the dialkylphosphoryl azides of Breslow et al. (1974).

3.3. Properties of diazo compounds

In Westheimer's original photoaffinity labeling experiment, diazoacetyl-chymotrypsin was photolysed. Later, Westheimer's group made the improved diazo reagents listed in Table 3.1.

3.3.1. Stability of diazo compounds

One of the major drawbacks of the diazoacetyl group was its instability at low pH and its reactivity in the dark towards protein functional groups (see Section 4.7.2). The 2-diazo-3,3,3-trifluoropropionyl and *p*-toluenesulfonyldiazoacetyl groups are considerably improved in these respects and are stable in 1 M hydrochloric acid.

Like aryl azides, diazo compounds are susceptible to reduction by thiols and a study of the diazotrifluoropropionyl group has been made by Takagaki et al. (1980) who showed that the corresponding hydrazone is formed (Fig. 3.6). For example, one reagent ~ 20 μM) was converted to the extent of 24% in 3 h at room temperature by 30 mM cysteine, at pH 8.0.

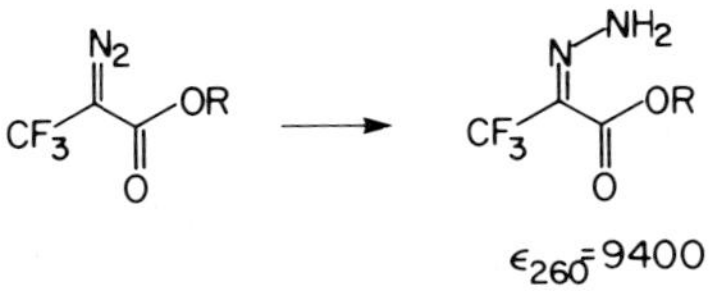

Fig. 3.6.

While azides were far more susceptible to reduction by dithiols, the rates of reduction of a diazotrifluoropropionyl derivative by dithiothreitol, β-mercaptoethanol, cysteine, and reduced glutathione did not differ widely. Thioglycolic acid was however a poor reductant and it was suggested that it should be used to replace β-mercaptoethanol or DTT when diazo reagents are used. The reduction may be monitored by TLC or by a ~ 500-fold increase in the absorbance at 260 nm.

The stability of a prospective diazo reagent in the dark should further be ensured by a careful consideration of its structural features. A problem that was encountered with N^6-diazomalonoyl cAMP was the Dimroth rearrangement (Fig. 3.7).

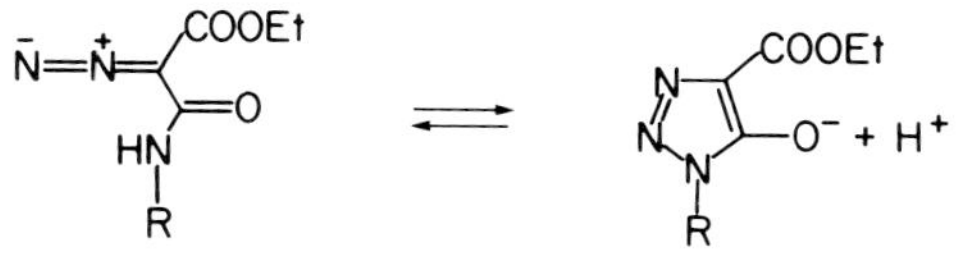

Fig. 3.7.

The equilibrium was to the right at higher pH values (> pH 7.0) and the triazole form was not efficiently photolysed (Brunswick and Cooperman, 1973; Guthrow et al., 1973). The Dimroth rearrangement is not always encountered with amides; for instance the diazomalonyl derivatives of aniline and benzylamine did not rearrange. It is not known whether diazotrifluoropropionamides and toluenesulfonyldiazoacetamides are susceptible.

A more subtle question concerns the two rotational isomers of diazo compounds with α-keto groups (Fig. 3.8).

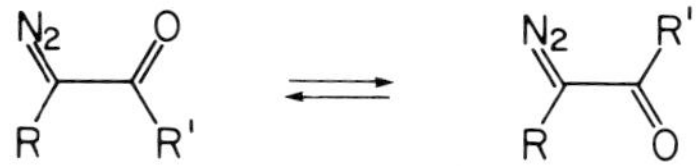

Fig. 3.8.

It has been suggested that the photochemistry of the two isomers may differ (e.g. Westheimer, 1980). In the case of the isomers of diazotrifluoropropionyl esters, interconversion is rapid at room temperature and it is possible that one or other of the isomers of a ligand could bind specifically to a receptor.

3.3.2. Size of diazoacyl derivatives

The synthesis of diazoacyl derivatives is undoubtedly more difficult (see below) than the synthesis of aryl azides. One strong advantage of the diazo derivatives may on occasion compensate for this inconvenience and that is the relatively small size of the functional group (Ganjian et al., 1978; Sen et al., 1982). The diazotrifluoropropionyl group is the smallest photoactivatable group developed so far that does not have severe drawbacks.

3.3.3. Photolysis of diazo compounds

The photolysis properties of most diazo compounds are not ideal for photoaffinity labeling. In general they possess a strong absorption band in the ultraviolet (Table 3.1) and a very weak band at longer wavelengths (e.g. $\varepsilon \sim 15$, at 340 nm). With the usual lamps irradiation at > 300 nm must be prolonged to achieve complete photolysis but Cooperman and Brunswick (1973) found that photolysis at 254 nm could be achieved in a reasonable time without damage to proteins if care was taken (see Chapter 4). Improvements in the absorption properties of diazo compounds were made when *p*-toluenesulfonyldiazoacetates (Chowdhry and Westheimer, 1978) and more recently dansyldiazomethylphosphinates (Stackhouse and Westheimer, 1981) were introduced (Table 3.1).

3.3.4. Undesirable side reactions of diazo compounds

Unfortunately not only carbenes are formed when diazoacyl and related compounds are photolysed. A major product is that formed by rearrangement of the carbene (or perhaps the excited diazo compound) to a ketene: the Wolff rearrangement (Fig. 3.9).

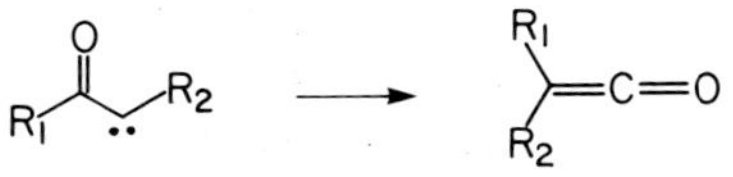

$$R_1C(=O)\ddot{C}R_2 \longrightarrow R_1(R_2)C{=}C{=}O$$

Fig. 3.9.

Ketenes are highly reactive electrophiles but not nearly so indiscriminate as carbenes. When the properties of diazoacetyl photoaffinity reagents were evaluated the Wolff rearrangement was found to be a major problem accounting for 30 to 60% of the products arising from *O*-esters and 100% of the products from *S*-esters. For instance, diazoacetyl-chymotrypsin gave rise to *O*-carboxymethyl serine formed by the attack of water on the ketene (Shafer et al., 1966) (Fig. 3.10).

Diazo compounds with electron withdrawing substituents were introduced both to improve the stability of the reagent in the dark (Table 3.1) and to

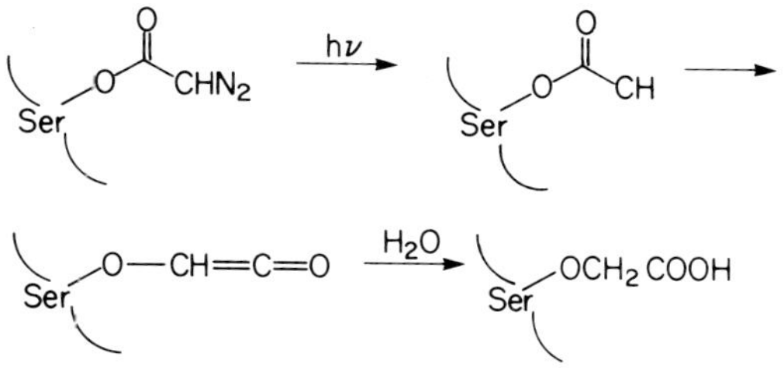

Ser–O–C(=O)–CHN_2 $\xrightarrow{h\nu}$ Ser–O–C(=O)–CH ⟶

Ser–O–CH=C=O $\xrightarrow{H_2O}$ Ser–OCH_2COOH

Fig. 3.10.

decrease the tendency toward Wolff rearrangement. (Reactivity in the dark was noted as a possible problem with diazoacetyl reagents when it was found that an acid catalysed reaction of diazoacetyl-chymotrypsin involving a histidine residue occurred in a control experiment, in the dark, at pH 6.2 [Shafer et al., 1966; and Section 4.7.2)]. When thiol esters of diazotrifluoropropionic acid are irradiated in methanol, 40% insertion into the –OH bond is found (none would have occurred with a simple diazoacetyl ester). While less insertion (25%) was observed with a thiol ester of toluenesulfonyl diazoacetic acid, such reagents have a greatly enhanced absorption band in the near ultraviolet (a broad absorption at 370 to 390 nm, ε 140; Chowdhry and Westheimer, 1978a,b).

3.3.5. Stability of the photoinduced crosslink

While little work has been done to determine the nature of the photochemical adducts between aryl nitrenes and protein receptors, Westheimer and his colleagues have put considerable effort into elucidating the structure of the products formed when carbenes react with proteins. The isolation, albeit in low yields, of insertion products formed by the reactions of carbenes with the –OH of tyrosine, the CH_3– of alanine and the –S_2– of cystine in chymotrypsin and trypsin augers well for the formation of stable products in other systems (see Chowdhry and Westheimer, 1979; and references therein).

3.3.6. Other diazo compounds as photoaffinity reagents

This section concludes by briefly considering the suitability of diazo compounds other than those listed in Table 3.1 as photoactivatable reagents. Among the possibilities are simple diazo alkanes, diazoketones, aryl diazo compounds, and diazophosphonates. Diazoalkanes are ruled out as reagents because of their extreme reactivity, particularly in the presence of proton donors. Besides, the carbenes derived from them would rearrange to olefins. Diazoketones are susceptible to the Wolff rearrangement, nevertheless they have been used in a number of photolabeling studies (for a list see Bayley and Knowles, 1977). In at least one case, however, attachment of the Wolff rearrangement product occurred at a position clearly outside the receptor site (Richards et al., 1974). Diazoketones are chemically reactive (Section 4.7.2). Most diazophosphonates are unstable in aqueous solution and are probably unsuitable as photoactivatable reagents (Goldstein et al., 1976; Bartlett and Long, 1977). Certain molecules may be of borderline utility, for example (Fig. 3.11).

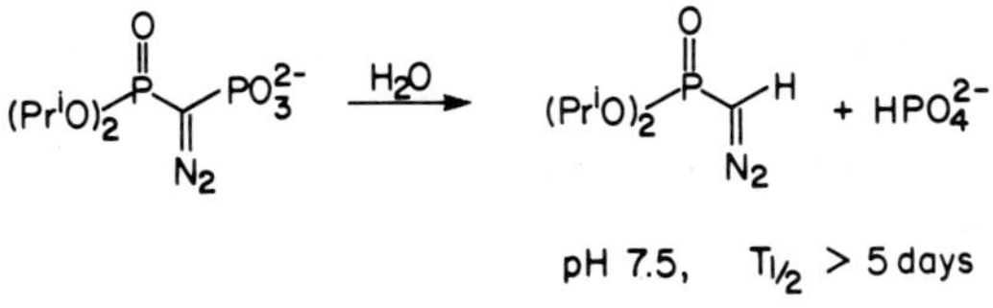

$$(Pr^iO)_2P(=O)C(=N_2)PO_3^{2-} \xrightarrow{H_2O} (Pr^iO)_2P(=O)C(=N_2)H + HPO_4^{2-}$$

pH 7.5, $T_{1/2} > 5$ days

Fig. 3.11.

Aryl diazomethanes vary in stability. They are formed when diazirines are irradiated and will be be discussed below.

3.4. Properties of diazirines

Diazirines, carbene precursors, have slowly gained in importance since Smith and Knowles (1973, 1975) first made 3-H-3-aryldiazirines and suggested that they would make useful reagents. Three useful classes of diazirines are given in Table 3.1.

3.4.1. Stability of diazirines

The three-membered ring is of unexpectedly high chemical and thermal stability, diazirines being far more stable than their linear diazo isomers. For instance, cyclohexanespirodiazirine (Fig. 3.14) is stable toward numerous highly reactive reagents: methyl iodide, peracetic acid, boron trifluoride etherate, bromine, common acids (except concentrated H_2SO_4), metal salts (Cu salts, Ni II, Pd II, Pt IV), triphenyl phosphine, lithium dimethylamide, diazomethane, acidic dichromate, silver oxide, acid chlorides, amines and alkaline hypobromite (Bradley et al., 1977). Such information is of great importance in designing synthetic strategies (see below). While aryl diazirines have not been bombarded with such an array of reagents they are clearly resistant to dilute acid and base, oxidising agents and mild reductants such as sodium borohydride. Diazirines are also stable to physiological conditions including the presence of thiols which attack azides and diazo compounds. Trifluoromethylphenyldiazirine was stable to 100 mM DTT at pH 9.8 for 24 h (Brunner et al., 1980).

While 3-H-3-phenyldiazirine was reported to be unstable on storage as a neat liquid (Smith and Knowles, 1975), its stability was improved in hexane solution at $-20°C$. Higher molecular weight derivatives of this compound have been stored for months in solution at $-80°C$ without decomposition.

3.4.2. The size of diazirine derivatives

Like aryl azides the useful diazirines are quite bulky but again they may often be built into molecules with structures similar to naturally occurring compounds (Fig. 3.12: Huang et al., 1982).

3.4.3. Photolysis of diazirines

The absorption properties of diazirines are characteristic. All possess an absorption band in the near ultraviolet (λ_{max} 350 to 380 nm) which is resolved into a series of sharp peaks in non-polar solvents such as hexane. Extinction coefficients are modest, usually around 300 M^{-1} cm^{-1}, but

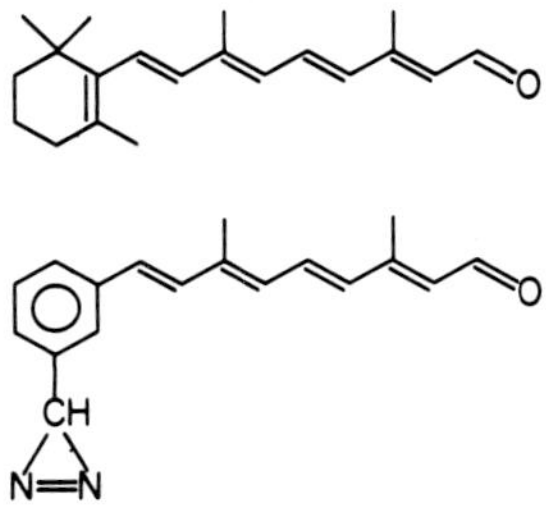

Fig. 3.12.

photolysis is efficient. As the N=N double bond in the diazirine system is not conjugated in aromatic derivatives, substituents on the aromatic nucleus have little effect on the absorption maximum (e.g. for 3-H-3-phenyl-diazirines: *p*-CH_3O-: λ_{max} 378; *m*-NO_2-: λ_{max} 352 nm).

3.4.4. Undesirable side-reactions of diazirines

Like diazo compounds and aryl azides, diazirines give rise in most instances to unwanted photolysis products besides the desired carbenes, including long-lived electrophilic intermediates. The most serious problem is the generation of diazo compounds (e.g. Fig. 3.13). For example, 3-H-3-aryl-diazirines form diazo isomers to the extent of 30 to 70% when irradiated. These isomers are themselves photolysed to form carbenes, but relatively slowly at wavelengths at which the diazirines absorb (Smith and Knowles, 1973, 1975).

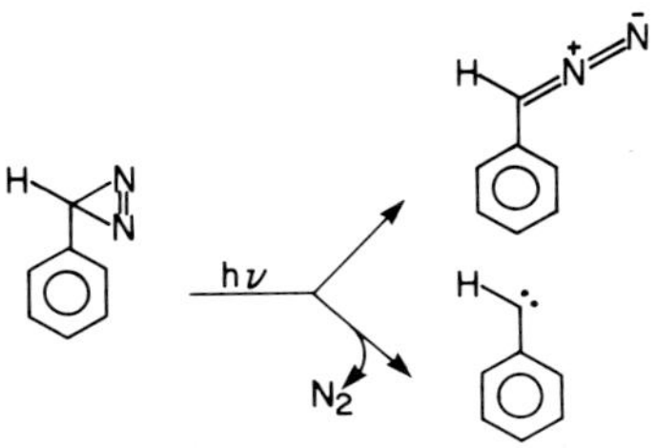

Fig. 3.13.

Diazoalkanes (such as 2-diazoadamantane) are highly reactive electrophiles. Phenyldiazomethane is less reactive but it has been used to modify proteins chemically (Delpierre and Fruton, 1965). Reaction with proteins in the dark after irradiating 3-H-3-phenyldiazirines has been noted (Ross et al., 1982) but it does not occur invariably (Standring and Knowles, 1980; Huang et al., 1982). Brunner and Richards (1980) were able to surmount the problem by introducing trifluoromethylphenyldiazirines. The linear diazo rearrangement products from these reagents are stable in 0.1 M CH_3COOH in cyclohexane for 24 h. Although the reactivity of these diazo compounds toward the various functional groups present in proteins and other macromolecules has not been tested systematically it seems that the trifluoromethylphenyldiazirines come closest of all the existing reagents to meeting the criteria listed at the beginning of this chapter.

The rearrangement of phenyl carbene to cyclohepta-1,2,4,6-tetraene has been detected at 10 K (West et al., 1982; cf. Fig. 3.3), but the relevance of this to photochemistry in solution at higher temperatures is not yet clear. Ortho-substituents that might react with photogenerated aryl carbenes should be avoided as they are with arylazides (Section 3.2.4; Fig. 3.4).

3.4.5. Stability of labeling products

A great deal is known about the photochemistry of carbenes (Chapter 2) and in the cases of the diazirine precursors given here (Table 3.1) it can confidently be predicted that the adducts formed on irradiation with proteins will be stable under most of the conditions necessary for further study (Fig. 2.1). Of course common sense should be exercized, e.g. insertion into carboxyl groups will yield esters which are unstable to hydrolysis or to treatment with hydroxylamine (e.g. Ross et al., 1982), and the possibility of reaction with peptide bonds culminating in polypeptide chain cleavage has already been mentioned (Section 2.1, Fig. 2.2).

3.4.6. Other diazirines as photoaffinity reagents

Just as we inquired for diazo compounds and azides, we may ask why certain diazirines are not used as photoaffinity reagents and what prospects there are for new diazirine reagents.

Unsubstituted alkyl diazirines are ruled out because carbenes from them rearrange to olefins (Fig. 3.14). The constraints of the caged ring system do not allow this possibility with the useful adamantane diazirine, although here a fraction of the carbene is lost by intramolecular insertion (Fig. 3.15).

Haloaryldiazirines can be synthesized in high yields but they should not be used as photogenerated reagents because some of their reaction products may be labile (for example, Fig. 3.16).

In an attempt to reduce the size of the photoreactive group, Erni and Khorana (1980) investigated the properties of tetrafluorodialkyl diazirines. Unfortunately these compact molecules (Fig. 3.17) did not yield insertion products on photolysis, but rearranged internally by alkyl migration to give olefins, and underwent intramolecular insertion to yield cyclopropanes.

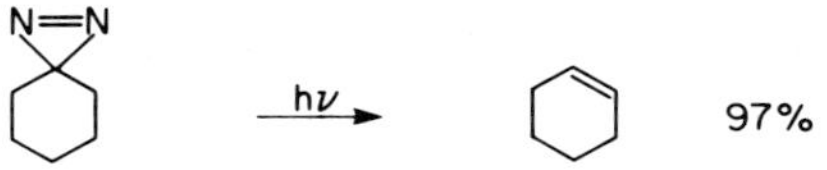

N=N hν 97%

Fig. 3.14.

Fig. 3.15.

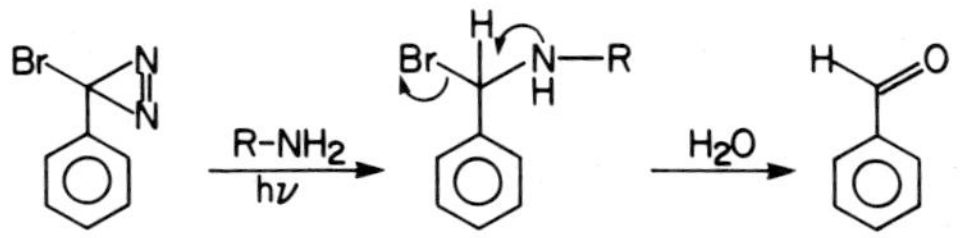

Br N N R-NH$_2$ hν Br H N—R H H$_2$O H O

Fig. 3.16.

N=N RCH$_2$CF$_2$ CF$_2$CH$_2$R

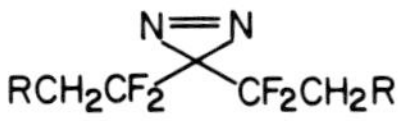

Fig. 3.17.

3.5. *Synthesis of photoaffinity reagents*

The discussion of the synthesis of photoaffinity reagents has been divided into two sections (3.5 and 3.6). First (Section 3.5), general methods for making azides, diazo compounds and diazirines are given. These methods are applicable to the direct synthesis of photoaffinity reagents, and to the synthesis of bifunctional reagents for attachment to other molecules. Bifunctional reagents, (which are also used for crosslinking: see Table 5.1) are most often used for derivatizing macromolecules for photoaffinity labeling (Section 3.6), but they are also used for constructing reagents by modifying small ligands. The latter strategy is popular as many small ligands are available in radioactive form. It is not discussed explicitly in Section 3.5 but some examples are given in Section 3.7 (on radiolabeling; see Fig. 3.27), and related topics, such as the functional group specificity of bifunctional reagents, are discussed in Section 3.6 and in Chapter 5.

In Section 3.5, only the most generally useful methods for making azides, diazo compounds and diazirines can be covered. In the unlikely event that these methods fail the reader may wish to try less commonly used alternatives. These may be found by searching for the syntheses of analogous compounds in *Chemical Abstracts*, or by consulting the appropriate volumes in 'Houben-Weyl: *Methoden der Organischen Chemie*', 'Beilstein', and 'The Chemistry of ...' series, e.g. *The Chemistry of the Azido Group*, S. Patai, ed., Interscience (1971).

Before a photoaffinity reagent is made, the literature on the relationship between the structure and function of the ligand should be consulted and the reagent designed accordingly. Where possible a reagent should first be made in non-radioactive form and tested for biological activity. It is one of the strengths of the photoaffinity labeling method that such assays can be carried out without the complication of covalent reaction (see Chapter 4).

3.5.1. *Synthesis of aryl azides*

The relative ease of synthesis of aryl azides has contributed greatly to their popularity as photoactivatable reagents. Two methods are generally used

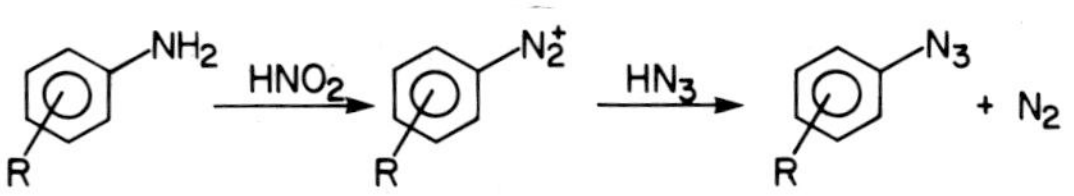

$$\text{R-C}_6\text{H}_4\text{-NH}_2 \xrightarrow{HNO_2} \text{R-C}_6\text{H}_4\text{-N}_2^+ \xrightarrow{HN_3} \text{R-C}_6\text{H}_4\text{-N}_3 + N_2$$

Fig. 3.18.

and the first, the conversion of aryl amines to azides by diazotisation followed by treatment with sodium azide, is the most common (Fig. 3.18).

Nitrous acid is generated in situ by the dropwise addition of aqueous sodium nitrite to a suspension or solution of the aromatic amine in a dilute inorganic acid (HCl, H_2SO_4 or HBF_4) at around $-5\,^\circ C$. Occasionally a lower temperature is required to prevent side reactions (e.g. Fleet et al., 1972). Usually the nitrite is added in slight excess and, if desired, a positive reaction on starch-iodide paper may be seen when no further amine is available for reaction. Excess nitrous acid is often destroyed with sulfamic acid or urea, or removed by degassing. After diazotisation a solution of sodium azide is added slowly and the aryl azide usually forms rapidly in high yield, often precipitating from the solution.

If necessary organic solvents such as acetic acid, ethanol, acetone, dimethylformamide, and dimethylsulfoxide are used to solubilize the reactants. In organic solvents, sodium nitrite may be replaced with isopentyl nitrite, and an azide salt with an organic counterion (e.g. triethylammonium) may be used. The reaction is generally carried out in a beaker as much frothing occurs as the N_2 evolves, and within a hood as HN_3 is highly toxic. In the case of recalcitrant amines large excesses of nitrite and azide must be used, e.g. Bastos (1975). By contrast, the mildest conditions must be sought for amines that also contain acid-sensitive functional groups such as esters.

Publications on the synthesis of aryl azides as photoactivatable reagents selected for their experimental procedures are: Fleet et al. (1972); Galardy et al. (1974); Katzenellenbogen et al. (1973); Schwyzer and Calviezel (1971); Bridges and Knowles (1974); Huang and Richards (1977). The reader should also consult the detailed description of Smith and Boyer (1963) and the *Methods in Enzymology* volume on Affinity Labeling

(Jakoby and Wilchek, 1977). The table of bifunctional reagents which will be discussed later (Table 5.1) contains numerous useful references.

A second method for making aryl azides is by nucleophilic aromatic substitution at positions highly activated by electron withdrawing groups. 1-Fluoro-2,4-dinitrobenzene, for instance, reacts rapidly with azide ion in DMF, at 20°C (Yoshioka et al., 1973). This reaction has been most important in the synthesis of 8-azidopurines as photoaffinity labels, notably by Haley and coworkers. For example, the synthesis of 8-azido-cGMP was recently described (Geahlen et al., 1979) (Fig. 3.19). For full experimental details of the synthesis of 8-azidoadenosine derivatives Czarniecki et al. (1979) should be consulted. 8-Azido-cIMP has also been made (Casnellie et al., 1978).

2-Azidoadenine derivatives are, incidentally, made by a different method, namely the diazotisation of the corresponding hydrazine (MacFarlane et al., 1982).

Aryl azides may be characterized by the usual methods of chemical analysis. Their ultraviolet and visible absorption properties described earlier, are distinctive. The weak shoulders at 280 to 290 nm are characteristic and are usually visible if they are not obscured by their major absorption band. For example, phenyl azide itself has λ_{max} 250 nm with shoulders at 277 and 286 nm. In the infrared, arylazides absorb strongly with a peak between 2,100 and 2,160 cm^{-1} which is often split due to Fermi resonance ($\Delta\nu = 50\ cm^{-1}$). Elemental microanalysis (for C, H, N) gives the expected values for pure aryl azides. Many examples of UV, IR, and NMR spectra of aryl azides are given in the literature cited.

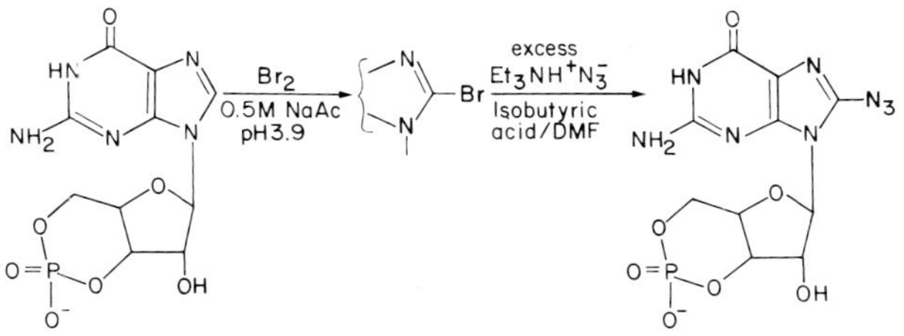

Fig. 3.19.

3.5.2. Synthesis of diazo reagents

Because the Wolff rearrangement limits their utility diazoacetyl reagents and diazoketones are now seldom used (but see, Ganjian et al., 1978; Sen et al., 1982). Useful guides to the literature on their syntheses have been given by Bayley and Knowles (1977) and by Sen (1982).

Diazomalonyl derivatives are occasionally used and descriptions of several useful reagents for introducing the ethyldiazomalonyl group have been given. Unlabeled ethyl diazomalonyl chloride may be made according to Hexter and Westheimer (1971) by reacting phosgene with excess ethyldiazoacetate (Fig. 3.20). The product should be distilled in a vacuum to remove reactive impurities including ethyl choroacetate and ethylchloromalonylchloride although this step has been omitted (Chakrabarti and Khorana, 1975). Ethyldiazomalonylchloride labeled with ^{14}C is usually made and used in situ. A vacuum line facilitates the handling of small amounts of the volatile labeled phosgene (Vaughan and Westheimer, 1969) but successful preparations have been carried out in a simple flask (Chakrabarti and Khorana, 1975; Ruoho and Kyte, 1974, 1977). Because the radiolabeled acylating agent is not purified the photoaffinity reagent that is made from it may be contaminated with the corresponding chloromalonyl derivative. Ethyldiazomalonylchloride has been used both to form amides by reaction with amines (Brunswick and Cooperman, 1971, 1973) and esters by reaction with alcohols (Ruoho and Kyte, 1974, 1977). Where a compound contains both a hydroxyl and an amino group the disubstituted derivative may be converted to the amide by treatment with 1 N NaOH (Brunswick and Cooperman, 1973).

The *p*-nitrophenyl ester of diazomalonic acid has been made. While this derivative proved useful for acylating the serine residues at the active sites of trypsin and chymotrypsin (e.g. Vaughan and Westheimer, 1969a,b;

$$2\,EtOOC\!-\!C(=N_2)\!-\!H + COCl_2 \longrightarrow EtOOC\!-\!C(=N_2)\!-\!C(=O)\!-\!Cl + EtOOCCH_2Cl + N_2$$

Fig. 3.20.

Hexter and Westheimer, 1971b) it is not a particularly 'active' ester and the *N*-hydroxysuccinimide ester, made from the acid choride, is better used for the acylation of amino groups (Bispink and Matthaei, 1973, 1977).

The synthesis of trifluorodiazopropionyl derivatives has been described by Chowdhry et al. (1976). The acid chloride, synthesized from 2,2,2-trifluorodiazoethane and phosgene, has been used to prepare esters from thiols and alcohols (Chowdhry et al., 1976). A description of the synthesis of the ^{14}C-labeled compound, its conversion to the *p*-nitrophenol ester and the acylation of chymotrypsin has also been given (Vaughan, 1970; Chowdhry et al., 1976). A similar synthesis of 2-diazo-2-*p*-toluenesulfonylacetylchloride and the *p*-nitrophenol ester derived from it has also been described; these reagents are indefinitely stable on storage at 25 °C (Chowdhry and Westheimer, 1978a,b). The acid chloride has been used to make thiol esters (Chowdhry and Westheimer, 1978a,b).

Recently, Stackhouse and Westheimer (1981) described the properties of (dansyldiazomethyl)methylphosphinic acid derivatives. They prepared *O*- and *S*-esters, and a phosphoramide by reactions of the acid chloride, formed in situ, in the presence of the catalyst dimethylaminopyridine (Fig. 3.21).

Using [^{3}H]dansyl chloride as the starting material, [^{3}H]dansyldiazomethane was synthesized in three steps at a specific radioactivity of up to

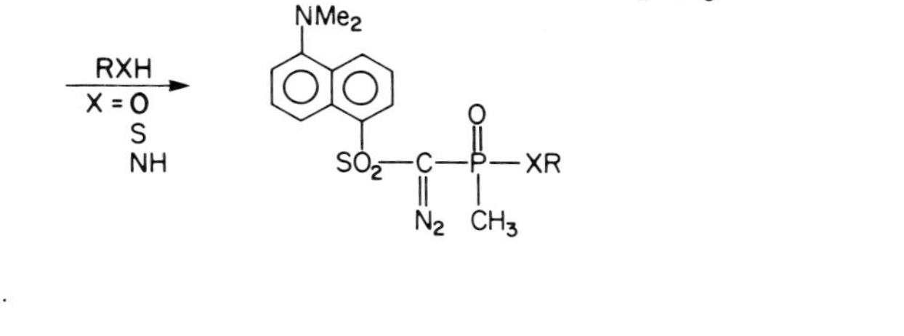

NMe2
SO2CHN2
CH3POCl2
N NMe2
NMe2
O
SO2 C P N NMe2
N2 CH3
RXH
X = O
S
NH
NMe2
O
SO2 C P XR
N2 CH3

Fig. 3.21.

37 Ci/mmol and subsequently converted to the methyl ester. The unlabeled *p*-nitrophenol ester of the phosphinic acid was also prepared.

Stable diazo compounds may be characterized by the usual means including NMR, absorption spectroscopy, IR spectroscopy, microanalysis, mass spectroscopy. Much excellent information may be found in the papers from Westheimer's group quoted here. The UV and visible absorption characteristics of the reagents of interest has already been given (Table 3.1). Like azides, diazo compounds have a strong absorption band in the IR: all the compounds discussed here absorb in the region 2,090 to 2,170 cm^{-1}

3.5.3. Synthesis of diazirines

Smith and Knowles (1973) made several 3-H-3-aryldiazirines and suggested that they had several advantages as photoaffinity labeling reagents over the aryl azides and diazo compounds that were then current. These diazirines did not become popular, however, because they are inconvenient to prepare. The situation today is unchanged.

In a thorough investigation, Smith and Knowles (1975) defined two routes for making 3-H-3-aryldiazirines. In the first an aldehyde is reacted with chloramine and ammonia to give a tricyclic compound that may be oxidized to a diazirine (Fig. 3.22). In the case of 3-H-3-phenyldiazirine the intermediate may be isolated and converted to the diazirine in 50% yield (Smith and Knowles, 1975; Bradley et al., 1977) but this failed in the case of other diazirines (p-CH_3-; p-CH_3O–) which were however prepared in

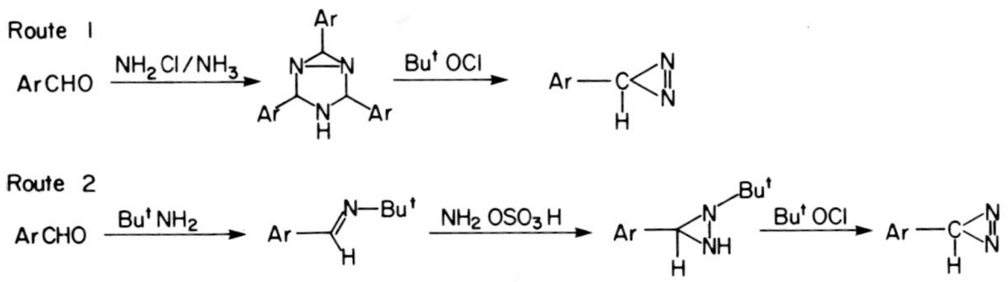

Route 1

ArCHO $\xrightarrow{NH_2Cl/NH_3}$ (tricyclic intermediate) $\xrightarrow{Bu^tOCl}$ Ar–CH(N=N)

Route 2

ArCHO $\xrightarrow{Bu^tNH_2}$ Ar–CH=N–Bu^t $\xrightarrow{NH_2OSO_3H}$ Ar–CH(N–Bu^t)(NH) $\xrightarrow{Bu^tOCl}$ Ar–CH(N=N)

Fig. 3.22.

very low yield (1 to 3.6%) by adding mercuric oxide during the formation of the triazabicyclohexane. This was said to trap a diaziridine intermediate. Radhakrishnan et al. (1981) have nevertheless successfully adapted the original method to prepare *m*-(methoxymethyleneoxy)phenyl-3H-diazirine, and other diazirines, in 10 to 15% yield from the aldehyde.

The second preparation also begins with an aldehyde and may be used to make 3-H-3-aryldiazirines with electron withdrawing substituents (Fig. 3.22). The sequence shown is not as forbidding as it appears as all the reactions are carried out in one pot. The yields are low, however (3 to 15%).

The low yields in these syntheses make purification of the diazirines a difficult task. Preparative TLC was used extensively by Smith and Knowles and this can now be effectively replaced by HPLC or flash chromatography.

In contrast to 3-H-3-aryldiazirines, adamantane diazirine may be prepared in high yield by a method (Isaev et al., 1973; Bayley and Knowles, 1980) based on a general procedure for dialkyl diazirines devised by Schmitz and Ohme (1961) (Fig. 3.23). Such a convenient synthesis is most attractive but as yet adamantane diazirines further functionalized to allow their attachment to molecules of biological interest, have not been prepared, although this should not be a difficult task.

Brunner et al. (1980) have reported on the properties of a new class of aryl diazirines, the 3-aryl-3-trifluoromethyldiazirines. Besides forming a class of the most attractive photoactivatable reagents available, these molecules may be made in far higher yield than the corresponding 3-H-diazirines. The route used, which incidentally fails with the 3-H-compounds (Smith and Knowles, 1975), is shown in Fig. 3.24. Two drawbacks are the large number of steps involved and the difficulty of preparing

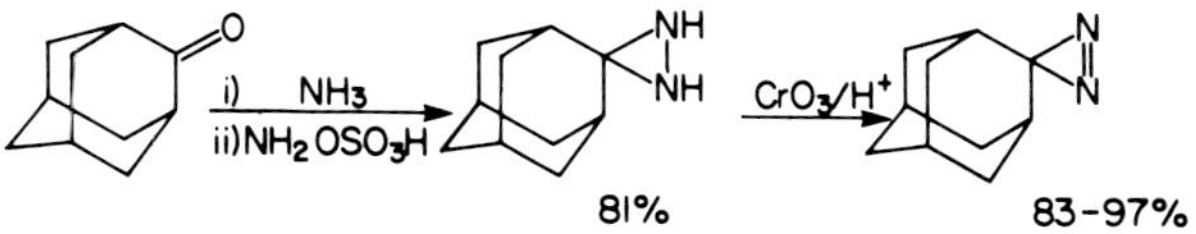

i) NH_3
ii) NH_2OSO_3H
81%
CrO_3/H^+
83-97%

Fig. 3.23.

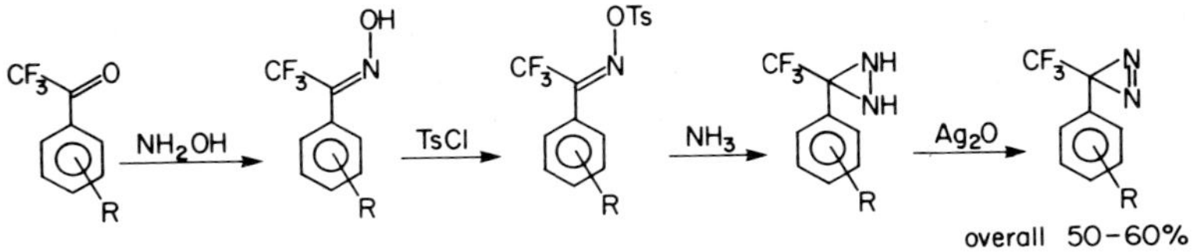

overall 50–60%

Fig. 3.24.

trifluoroacetophenones. Brunner et al. described the synthesis of a tosylate (Fig. 3.25), useful for further chemical elaboration (see, e.g. Brunner and Richards, 1980). Such useful synthetic intermediates have not been described for 3-H-aryldiazirines or adamantane diazirines although Smith and Knowles (1975) described a courageous preparation (1 % yield) of *p*-carboxymethoxyphenyl-3-H-diazirine which could be activated for attachment to proteins.

The distinctive absorption of diazirines in the near ultraviolet has already been described (Table 3.1) and this feature is most useful in making a preliminary estimate of the yield of a diazirine synthesis, in following the purification of the product, and in confirming its identity. Diazirines have an absorption of moderate strength in the IR between 1,540 to 1,595 cm^{-1} which has been reported as a single peak for 3-H-3-aryldiazirines (Smith and Knowles, 1975) and a doublet for dialkyl diazirines (Bradley et al., 1977). Unfortunately several other functionalities absorb in this region and the characteristic ultraviolet absorption remains the diagnostic test for diazirines. 3-H-3-aryldiazirines are somewhat unstable and may not always give satisfactory elemental analyses. The 3-H proton of the 3-H-3-aryldiazirines appears as a characteristic singlet at 1.75 to 2.18 δ depending on the aryl ring substituents (Smith and Knowles, 1975). ^{19}F NMR should be considered as an analytical method for the trifluoromethylphenyl diazirines.

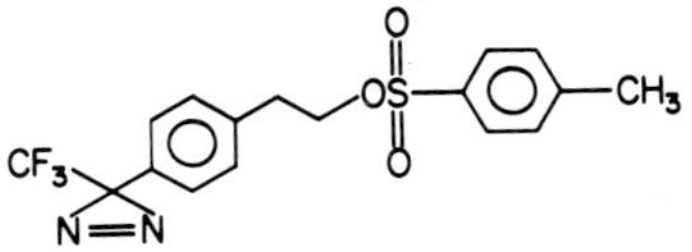

Fig. 3.25.

3.6. Synthesis of reagents by chemical modification of macromolecules

3.6.1. General considerations

Photoaffinity reagents of low molecular weight can be made by total synthesis or by attaching photoactivatable groups to preexisting ligands. For macromolecules, such as polypeptides, only the latter strategy is feasible.

If a conventional affinity reagent is prepared from a macromolecule, by attaching a chemical bifunctional reagent through one of its arms, there is the danger that the second reactive group will form intra- or intermolecular crosslinks before affinity labeling can be carried out. Only in cases where the macromolecular ligand of interest is lacking a particular functional group will such an experiment be straightforward. The problem can be circumvented by using a bifunctional reagent with one chemically reactive group, for attachment to the macromolecule, and one photoactivatable group, for crosslinking to the receptor. Many of the bifunctional reagents that have been used to derivatize macromolecules for photoaffinity labeling are listed in Table 5.1.

I shall focus on the derivatization of polypeptides and proteins for photoaffinity labeling. Other macromolecules including tRNAs (see, e.g. Schwartz and Ofengand, 1978) and oligonucleotides (Ivanovskaya et al., 1979) have been converted to photoactivatable derivatives and the principles are the same. Unmodified nucleic acids have also been used as macromolecular reagents (see, Section 2.4.4 for reviews, and for recent examples: Maly et al., 1980; Sperling et al., 1980; Park et al., 1982a,b; Cao and Sung, 1982), as have psoralen-DNA monoadducts (Wollenzien and Cantor, 1982).

The choice of a bifunctional reagent for modifying a protein requires the consideration of several important variables:

(i) *The photoactivatable group.* The pros and cons of the various photochemical reagents have been given above. Actually, the choice of available bifunctional reagents is largely limited to aryl azides. It is evident from the compilation in Table 5.1 that the ease of synthesis of aryl azides has stimulated a small industry for the production of new reagents.

(ii) *The chemically reactive group.* Azide derivatives designed for attachment to almost all the reactive groups in polypeptides have been made in the last five years. The choice of carbene precursors on the other hand is limited (Table 5.1).

(iii) *The dimensions of the reagent.* Chemical affinity labeling reagents were divided into two classes by Baker (1964). Endo reagents react covalently with functional groups at the active site of an enzyme (or at the heart of a ligand binding site in a receptor). Exo reagents bind at the active site but the reactive group, which is on the periphery of the molecule or on a side arm, reacts outside the active site. Occasionally, receptors labeled with ext reagents will still bind ligand (Lawson and Schramm, 1962; Silman and Karlin, 1969; and Fig. 4.6). For reagents of low molecular weight the distinction between endo- and exo-labeling often becomes a matter of semantics, but with macromolecular reagents it deserves serious consideration, although no systematic study of the problem has yet been made.

Presumably, a polypeptide ligand modified near the region which binds to the receptor will be inactivated as most of the bifunctional reagents (Table 5.1) are bulky, and on chemical attachment they often neutralize

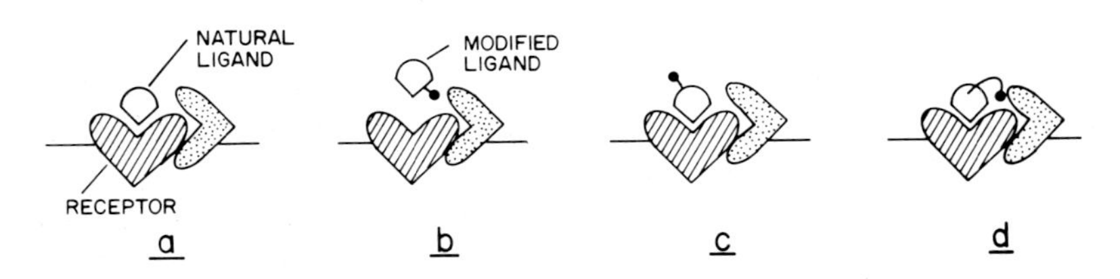

Fig. 3.26. The interaction of macromolecular photoaffinity reagents with a receptor. *a:* The natural ligand binds tightly to a receptor on a membrane (or in solution). *b:* A ligand modified with a bifunctional reagent close to its site of interaction with the receptor may not bind well. *c:* If the ligand is modified with a short-armed reagent at an alternative site it may not label the receptor but, *d:* if a long-armed reagent is used neighboring or irrelevant polypeptides may react with the reagent.

negatively or positively charged groups. However, if a short-armed reagent is appended at a point distant from the region involved in binding, it may label nothing (i.e. solvent) or irrelevant proteins. A long-armed reagent while increasing, up to a point, the chances of labeling the receptor will also increase the likelihood of irrelevant labeling (Fig. 3.26). The conclusion is that, to obtain unequivocal results the investigator should be prepared to construct several reagents. Bifunctional reagents should be attached at different points on the macromolecule.

In the absence of detailed structural information the point of attachment might be varied by using reagents with different functional group specificities.

It is worth noting that long hydrocarbon linkers may coil in aqueous solution and hydrophilic arms are preferable.

(iv) *Cleavable side-arm.* A cleavable side-arm is used in the true cross-linking experiments described in Chapter 5. If radiolabel is incorporated at the same end of the molecule as the photoreactive group, radioactivity may be transferred from a macromolecular reagent to its receptor by cleavage of the crosslink, facilitating the identification of the receptor (e.g. Maassen, 1979; Schwartz et al., 1982).

(v) *Other considerations.* Other considerations include the solubility of the reagent in the buffer in which the protein is derivatized. Several reagents (Table 5.1) are rather insoluble and the addition of organic solvents may be required to solubilize them for reaction. If this is the case, care must be taken not to irreversibly denature the protein under investigation. The availability of radioactive reagent is a further consideration. Often, however, the protein is labeled and modified in separate reactions (see Section 3.7). Finally the perturbation to the structure of the protein must be considered. For instance, to avoid altering the charge of a protein on derivatization of an amino group, imidates have been used to yield amidines which are protonated at physiological pH values.

3.6.2. *Some examples of syntheses of photoactivatable polypeptides*

Both rational and 'shot-in-the-dark' approaches have been used for preparing photolabile polypeptides. In the former, an attempt is made to modify a specific residue of a protein or peptide, or to modify the molecule and then identify the site of modification. This approach is commendable as an uncharacterized reagent that might be a mixture of active, partly active and inactive derivatives is more likely to give equivocal results than a pure, characterized molecule. Of course, this approach is only practicable when the protein chemistry of the polypeptide under study has been previously explored. The second approach is used with ill-characterized polypeptides or perhaps for pragmatic reasons with those whose structure is understood. The polypeptide is reacted with sufficient reagent to modify each molecule at least once, and the nature of the derivative(s) is not investigated except to demonstrate its biological activity.

Some representative examples of polypeptide and peptide modification with photolabile reagents are now given to illustrate the principles discussed above. Water-soluble reagents are convenient to use. For example, an arylazido derivative of radioiodinated calmodulin has been prepared by reaction with methyl 4-azidobenzimidate in borate buffer at pH 9.8 (Andreasen et al., 1981; Hinds and Andreasen, 1981). In this case the product was not fully characterized but it was noted that when more than one azido group was attached to the protein its ability to stimulate phosphodiesterase was diminished. When a poorly soluble reagent is used it must be added in an organic solvent that is miscible with water (e.g. DMF, DMSO, dioxane, ethanol). The activities of many polypeptides are unaffected by low amounts ($< 10\%$ v/v) of these solvents. Lewis et al. (1977) used the *N*-hydroxysuccimide ester of 2-nitro-5-azidobenzoic acid to derivatize phospholipase A_2. The reagent was added in dioxane, the final concentration of which was 10%. After 4 h at room temperature the modified protein was separated from low molecular weight contaminants by gel filtration. Beneski and Catterall (1980) used the same procedure to prepare a derivative of radiolabeled scorpion toxin for labeling sodium channel constituents. Unlabeled bovine serum albumin was present in the reaction mixture as carrier and after quenching the reaction with a buffer containing Tris, the mixture was used directly for photoaffinity labeling. Other cases where modifying reagents were added in organic solvents include the reaction of cytochrome *c* with 2-nitro-4-azidophenyl fluoride (0.1 M $NaHCO_3$, pH 9.5 containing 7% ethanol: Bisson et al., 1978) and with 2,4-dinitro-5-azidophenyl fluoride (aq. Na_2CO_3 with 10% ethanol: Erecinska et al., 1975). In some cases, where the protein is easily renatured derivatization may be carried out entirely in an organic medium. Levy (1973) reacted insulin with a large excess of 2-nitro-4-azidophenyl fluoride in DMF containing triethylamine. After 5 h at room temperature the protein products were precipitated by the addition of ether before chromatography on DEAE-Sephadex. Yip et al. (1980) also used DMF as a medium for reacting insulin with the *N*-hydroxysuccinimide ester of 4-azidobenzoic acid. Often, the concentration of organic solvent can be reduced to surprisingly low levels, e.g. Schwartz et al. (1982) reacted gelatin and fibronectin with 3-[(2-nitro-4-azidophenyl)-2-aminoethyldithio]-*N*-succinimidyl propionate in phosphate buffered saline containing as little as 0.5% DMF.

With small peptides the problem of denaturation does not arise and, further, there is often available a unique position for derivatization providing an opportunity for the rational synthesis of a derivative. For example, the 39 residue corticotropin was specifically derivatized at residue 9, the unique tryptophan, with nitroazidophenylsulfenyl chlorides (Canova-Davis and Ramachandran, 1980; Muramoto and Ramachandran, 1980). The reaction was done in 90% acetic acid with a large excess of methionine present in the mixture to act as a scavenger, protecting the methionine residue of corticotropin from oxidation. Stadel et al. (1978) derivatized the N-terminal amino group of oxytocin by reaction with the *N*-hydroxysuccinide ester of *N*-2-nitro-5-azidobenzoylglycine in pyridine. Niedel et al. (1980) reacted 2-nitro-4-azidophenyl fluoride with the terminal lysine of the chemotactic peptide Nle-Leu-Phe-Nle-Tyr-Lys. In a recent development, Gorman and Folk (1980) have used transglutaminase to catalyse the attachment of photolabile amines to peptides. For example, residue 5 of substance P (11 residues, glutamines at positions 5 and 6) was specifically modified with *N*-(2-nitro-4-azidophenyl)ethylenediamine.

Peptide semi-synthesis has also been used to produce photolabile peptides. Lee et al. (1979) coupled *N*-(2-nitro-4-azidophenyl)ethylenediamine to the carboxyl group of Boc-Tyr using a carbodiimide. As 100% CF_3COOH destroyed the azide group, the protecting group was removed with dilute CF_3COOH in methylene chloride/acetonitrile in the presence of anisole, a carbonium ion scavenger. The deprotected tyrosine derivative was reacted with Boc-Tyr-D-Ala-Gly-Phe-Met and carbodiimide to yield, after deprotection, an enkephalin analog.

In a more elaborate example, Thamm et al. (1980) selectively removed, by Edman degradation, the N-terminal glycine of the A-chain of an insulin derivative in which the B-chain amino terminus and the B-chain lysine were protected. The new N-terminus was reacted with the *N*-hydroxysuccinimide ester of *N*-(4-azido-2-nitrophenyl)glycine to yield, after deprotection, a derivative of insulin modified only at the $N^{\alpha A}$-terminus. The amino protecting group chosen for this procedure was the base labile, methylsulfonylethyloxycarbonyl.

3.6.3. Characterization and purification of macromolecular reagents

The first step in characterizing a modified peptide or protein is to estimate the number of photoactivatable groups incorporated per molecule. This is not just to satisfy the curiosity of the investigator: the number of modified residues should be recorded as it may be strongly correlated with the activity of the polypeptide derivative (e.g. Andreason et al., 1981) or even with its solubility (Erecinska, 1977). As the most satisfactory reagents absorb at higher wavelengths than the polypeptide itself or at least very strongly in the 280 nm region, the extent of modification can usually be estimated by absorption spectroscopy (for examples see: Ji, 1977; Levy, 1973; Andreason et al., 1981; Bisson et al., 1978; Randolph and Allison, 1978; Erecinska, 1977). Witzemann et al. (1979) designed aryl azides based on Ellman's reagent for coupling to thiol groups. The extent of modification could be followed by monitoring the released 5-thio-2-nitrobenzoic acid anion at 412 nm after adjusting a portion of the reaction mixture to pH 7.0. Alternative approaches include amino acid analysis of the derivatized protein (e.g. Nathanson and Hall, 1980). Of course, if a radiolabeled reagent has been used, estimation of the extent of derivatization is simple.

In many cases the derivatized protein is used directly in photoaffinity labeling experiments. Such preparations are often mixtures of several, or even numerous derivatives of the protein, but disregarding aesthetic considerations, such mixtures have been used successfully to identify receptors. Nevertheless, nightmarish situations can be postulated where, for example, a tightly binding component of the mixture which labels with poor efficiency (this might even be underivatized protein) blocks the receptor to access by a more weakly binding component which is, however, a good labeling reagent (cf. Fig. 3.26)*.

A homogeneous reagent can often be obtained by fractionating the derivatized protein. This task is made simpler if the number of possible products is limited. Epidermal growth factor (MW 6,000) contains a single amino group at the N-terminus and so it can be assumed that a single product was obtained on reaction with an arylazido imidate (Das et al., 1977). Insulin on the other hand contains three available amino groups and its derivatives may be separated by ion-exchange chromatography (Levy, 1973; Yip et al., 1980). A trisubstituted derivative was found to be inactive (Yip et al., 1980). While the monosubstituted $N^{\varepsilon B29}$-compound could be obtained directly by reaction with the *N*-hydroxysuccinmide ester of 4-azidobenzoic acid, preparation of the $N^{\alpha B1}$-derivative was facilitated by selective protection of the two other amino groups (Yeung et al., 1980). Bisson et al. (1978, 1980) reacted cytochrome *c* with 2-nitro-4-azidophenyl fluoride and obtained two monosubstituted species after ion exchange chromatography. One, the Lys-13 derivative, labeled subunit II of cytochrome oxidase while the second, the Lys-22 (?) derivative, did not label any protein components of the oxidase. After using other bifunctional reagents it has proved difficult to obtain homogeneous derivatives of

* We are assuming that a highly purified protein was used to begin with. In some cases, this is extremely important as a biologically active contaminant might be derivatized which efficiently labels a different receptor to that being sought. A protection experiment (Section 4.7.3) using the contaminated preparation would not reveal this problem.

cytochrome *c* even by ion-exchange chromatography (Erecinska et al., 1975; Erecinska, 1977; Randolph and Allison, 1978).

3.6.4. Total synthesis of small peptides

Using photolabile amino acids as building blocks, small photolabile peptides can be made by total synthesis, rather than by chemical modification. This method is preferable as analogs with structures closely similar to the unmodified ligand can be made. However, total synthesis will not become a method of choice in most laboratories until reagents fully compatible with rapid solid-phase peptide chemistry have been fully explored.

Much of the work in the area of photoactivatable peptides has been done by Schwyzer, Escher and their coworkers. L-Amino acids that have been used in peptide synthesis include 4′-azidophenylalanine (Schwyzer and Calviezel, 1971), 3′-azidophenylalanine (Escher and Schwyzer, 1975), 4′-nitrophenylalanine (Escher and Schwyzer, 1974, 1975) and 2′-nitro-4′-azidophenylalanine (Escher and Schwyzer, 1975; Fahrenholz and Schimmack, 1975; Staros and Knowles, 1978). *p*-Trifluoromethyldiazirinophenylalanine has recently been synthesized (Shih and Bayley, unpublished). Exemplary syntheses involving photoactivatable amino acids include an α-melanotropin analog (Eberle and Schwyzer, 1976) and an angiotensin II analog (Escher et al., 1978). Fischli et al. (1976) made L-4′-azido-3′,5′-ditritiophenylalanine from L-4′-amino-3′,5′-diiodophenylalanine by catalytic reduction with tritium gas followed by diazotization and treatment with sodium azide. In the syntheses cited above it was demonstrated that the amino group of 4′-amino-3′,5′-diiodophenylalanine is so hindered that it will not readily undergo acylation. It may be used unprotected and converted to the tritiated azide part way through or at the end of a synthesis. A full range of protecting groups has not been explored for compatibility with photolabile amino acids. The azido compounds appear to be compatible with the protection of amino groups by *t*-butoxycarbonyl (e.g. Eberle and Schwyzer, 1976; Staros and Knowles, 1978; Lee et al., 1979), methylsulfonylethyloxycarbonyl (Thamm et al., 1980) and trifluoroacetyl (Thamm et al., 1980), and *t*-butyl ester protection of glutamic acid (Eberle and Schwyzer, 1976). 4′-Nitrophenylalanine is resistant to HF, which is often used to cleave peptides from supports in solid phase synthesis, while 4′-azidophenylalanine is rapidly decomposed (Escher, 1977).

3.7. Radiolabeled reagents

It may seem obvious that specific radioactivity should be a primary consideration in the design of a radiolabeled reagent, yet it is surprising how often this aspect is poorly considered. Many receptors are present in extremely low concentrations in biological preparations effectively ruling out the use of long-lived isotopes which have low maximal specific activities (e.g. ^{14}C : 62.4 mCi/mA) compared with relatively short-lived isotopes such as ^{3}H (29,000 mCi/mA) and ^{125}I (2,170,000 mCi/mA). An estimate must always be made of the expected extent of label incorporation (assuming a modest labeling efficiency, say 5 to 10%) to determine whether it will be sufficient for the analytical techniques that will be employed. For example, if a receptor were present at a low level in a membrane preparation, say at 10 pmol/mg protein, and it were labeled at 10% efficiency with a ligand of specific activity 1 mCi/mmol, only 0.22 dpm would be present in a labeled band in an SDS-polyacrylamide gel to which 100δ g protein had been applied: insufficient for detection.

A second question is how and where to place the radiolabel in the reagent. Three strategies can be identified: (i) A direct synthesis of a photoaffinity reagent, designed from first principles, is performed, using a radioactive reagent in one of the steps. (ii) The photoaffinity reagent is formed either by attaching a photoactivatable group to an available radiolabeled ligand, or by attaching a radiolabeled, photoactivatable group to an unlabeled ligand. This has the advantage that the radiolabeled components may be commercially available. (iii) An unlabeled molecule is modified twice, once to introduce the photoactivatable group, and again to introduce radiolabel. This strategy is often used with macromolecules.

Obviously the circumstances are so varied that only the major possibilities can be considered here. The volumes by Murray and Williams (1958) remain a useful source of synthetic methods for small molecules. The major radiochemical suppliers now stock an impressive array of reagents useful for radiochemical synthesis including $^3H_2$, NaB^3H_4, ^3H_2O, $K^{14}CN$, $Na^{125}I$, [^{32}P]phosphate, [^{14}C]carbonate, [^{3}H]- or [^{14}C]acetic anhydride and methyl iodide, as well as more exotic molecules. The radiochemical companies provide tritium labeling services for reactions involving tritiated water, tritium gas, borohydrides or methyl iodide. The services for tritium gas, which is difficult to handle, and tritiated water, which is not commercially available at the highest specific radioactivities, are particularly valuable. Tritium from water or from tritium gas can be incorporated into organic molecules by exchange in the presence of a catalyst such as palladium on charcoal. Photolabile molecules have been directly subjected to such processes (e.g. Witzemann and Raftery, 1978). However, as most photoaffinity reagents contain easily reduced groups, it would be prudent to consider tritiation of a precursor rather than the reagent itself, or to use acid or base catalysis. In the case of an aryl azide, the amine precursor may

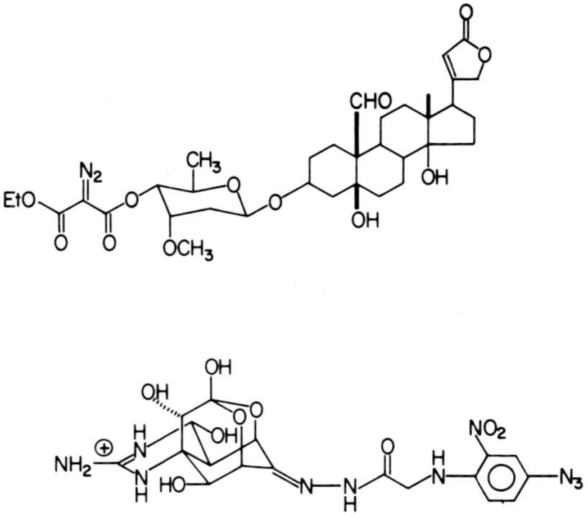

Fig. 3.27. Photoaffinity reagents made by attaching photoactivatable groups to readily available ligands. *a:* Cardiac glycoside derivative for reaction with Na,K-ATPase. The reactive secondary hydroxyl of cymarin was reacted with ethyldiazomalonyl chloride (Ruoho and Kyte, 1974, 1977). *b:* Tetrodotoxin derivative for reaction with the voltage-sensitive Na-channel. Tetrodotoxin was oxidised with periodate to form a ketone which was reacted with an azidoarylhydrazide (Chicheportiche et al., 1979).

be tritiated by an exchange procedure (examples are: Hanstein et al., 1979; Marinetti et al., 1979; Amitai et al., 1982). Hydrogenolysis of an alkyl or aryl halide under basic conditions is a useful alternative to the exchange procedures (Fischli et al., 1976; Bayley and Knowles, 1980). Diazirines resist $NaBH_4$ and molecules containing this functionality have been labeled by reducing carbonyl groups with borotritide (Huang et al., 1982).

In the above procedures it is advisable to place the radiolabel as close as possible to the photolabile group. If the two are separated by a labile linkage such as an ester group, problems can arise if anything but the mildest conditions are used during the analysis of the labeled molecules. When ^{32}P-labeled nucleotides are used the position of labeling must again be carefully considered. For example, enzymatic transfer can occur from [γ-^{32}P]8-N_3ATP, and [α-^{32}P]8-N_3ATP may be the most appropriate reagent in some experiments, although revealing information can sometimes be obtained by using both reagents (Potter and Haley, 1982).

Where a photoreactive group is attached to a preexisting ligand (a useful method for making reagents; Fig. 3.27) the same principle for placement of the radiolabel should be followed, and a useful formula for constructing radiolabeled bifunctional reagents for attachment (or for crosslinking experiments; Chapter 5) is to join in series:

(i) The photolabile group; (ii) a commercially available radioactive linker (glycine in the example); and (iii) a chemically reactive group (Fig. 3.28). Of course, many ligands are themselves available in radioactive form and the synthesis of a photoaffinity reagent may simply entail attaching a photoactivable group using, for example, one of the molecules listed in Table 5.1 (Fig. 3.27).

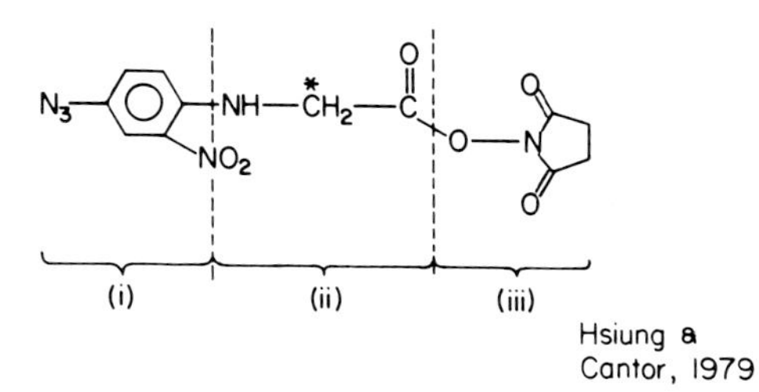

Fig. 3.28.

The construction of radiolabeled macromolecular photoaffinity reagents usually falls into the third category given at the beginning of this section in which the photoactivatable group and the radiolabel are attached in separate steps, although there is no reason why the photoactivatable group and radiolabel should not be introduced together, indeed this may be preferable (Schwartz et al., 1982). Radioiodination has successfully been used in the preparation of photoactivatable macromolecules. The iodination, often done by the chloramine-T method, may be carried out before (e.g. Beneski and Catterall, 1980) or after (e.g. Yeung et al., 1980) attachment of the photolabile group (for an important caveat concerning the chloramine-T method see Comens et al., 1982; and Section 4.7.2). The 2-nitro-4-azidophenyl group was unaffected by the chloramine-T procedure (Levy, 1973). Often, not all iodinated species are active, and diligent workers have occasionally taken the time to separate the active molecules before attachment of the photolabile group (e.g. Beneski and Catterall, 1980). Polypeptides may also be radiolabeled with Bolton-Hunter reagent (*N*-succinimidyl-3-(4-hydroxy-5-[^{125}I]iodophenyl)propionate) or by reductive alkylation. If the photoactivatable group is to be attached to a lysine residue it should be kept in mind that complications might arise because amino groups will already have been blocked by the above procedures.

The storage of radiochemicals has been studied extensively (Sheppard, 1972; Evans, 1976; much useful information is found in the product catalogs of the major radiochemical suppliers). In general, long-term storage should be at the lowest temperature possible, preferably in the absence of oxygen, in a hydroxylic solvent (e.g. ethanol or water containing 1 to 5% [v/v] ethanol [or if possible β-mercaptoethanol] as a radical scavenger) at below 1 mCi/ml. Toluene is a useful solvent for hydrophobic molecules. Unless liquid nitrogen temperatures are available for storage, samples should not be frozen as this often leads to the formation of microcrystals in which radiolytic decomposition rapidly occurs at the usual freezer temperatures. Crystal formation may be minimized by very rapid freezing.

In a dramatic example of radiochemical instability Bernier and Escher (1980) found that an analog of angiotensin containing 4′-azido-3′,5′-ditritiophenylalanine (73 Ci/mmol^{-1}) decomposed completely in one day at

$-200\,^{\circ}$C when dry. At $-200\,^{\circ}$C, in water containing 4′-aminophenylalanine as a scavenger the hormone analog was stable for more than a year.

The stability of the radiolabel towards ultraviolet irradiation must also be considered. With radioiodine this is particularly important: aryl iodides have themselves been used as photoaffinity reagents based on the homolytic cleavage of the C-I bond to form radicals (Section 2.3.2). Demoliou and Epand (1980) irradiated an iodinated photoactivatable derivative of glucagon, and noted a loss of ^{125}I that was more rapid at pH values above the pK_a of iodotyrosine (p$K_a = 8.2$). In this and in numerous other examples it has been possible to achieve useful extents of labeling by keeping the irradiation time to the minimum required for activation of the photolabile group, by irradiating at the longest possible wavelength (λ_{max} iodotyrosine anion = 305 mm), and the lowest possible pH values.

Lability of the radiolabel during photolysis occurs in other types of molecules and the investigator should be aware of the peculiarities of those that he deals with. For example, ^{3}H can be lost from photohydrates of pyrimidines (Wang, 1976).

CHAPTER 4

The photoaffinity labeling experiment

Photoaffinity labeling experiments are discussed in this chapter. Many of the techniques given in detail here are also applicable to other experiments such as photochemical crosslinking (Chapter 5) and studies on membrane topography (Chapter 6). The ultimate goal of a photoaffinity labeling experiment should be kept in mind; it is either to label and identify a macromolecular receptor in a complex mixture of biological molecules, or to identify the region within a macromolecule responsible for binding a ligand. It is now assumed that the reader possesses (Chapter 3), in his imagination at least, a photoactivatable ligand with which to proceed with the labeling experiments.

4.1. Biological activity of the photoaffinity reagent

Unlike conventional affinity labeling reagents, photoaffinity reagents may be assayed for biochemical or biological activity, in the dark, with none of the complications that arise with chemically reactive substances. These assays may often be carried out with the unlabeled reagent, and the investigator should be prepared to study the properties of a range of ligands, before embarking on a radiochemical synthesis, with the aim of identifying one or more that bind tightly to the receptor. It is most important that the photoaffinity reagent binds tightly (see Section 4.2) and the opportunity to determine the binding constant (i.e. K_d, K_i or K_m) must be taken.

As photoaffinity reagents can be used in the same manner as the natural ligands conventional methods are used to determine their properties and it will suffice to give a few examples and some caveats here.

Enzymatic assays have proved useful in defining the properties of numerous photoaffinity reagents. Occasionally, the reagents themselves have been the enzyme substrates. For example, glycyl(2′-nitro-4′-azido)-L-phenylalanine was shown to be a substrate for the dipeptide transport system of *E. coli* (Staros and Knowles, 1978), and photoactivatable derivatives of cytochrome *c* that restore O_2 uptake to cytochrome *c* depleted mitochondria have been prepared (for example Erecinska, 1977). Recently, DeRiemer and Meares (1981) prepared β-(4-azidophenyl)adenyl(3′,5′)uridine 5′-diphosphate. The dinucleotide was a substrate for RNA polymerase and photoaffinity labeling was carried out before and after extending the probe by one or two more nucleotides by the action of the enzyme. The subunits of the enzyme in contact with the 5′-end of the growing RNA chain could thus be determined. Clearly such an experiment can only be done with a photoactivatable molecule that is a good substrate.

More often the photoaffinity reagent is a competitive or non-competitive inhibitor of a biological activity. Kaczorowski et al. (1980) labeled the lactose transporter of *E. coli* by photochemical nucleophilic aromatic substitution using 4-nitrophenyl-α-D-galactopyranoside which had earlier been shown to a potent competitive inhibitor of lactose uptake. Brems and Rilling (1979) synthesized several aryl and alkyl azido pyrophosphates designed to bind at the active site of phenyltransferase. One of them, *o*-azidophenethyl pyrophosphate had a low K_i in a kinetic assay and was later used in successful labeling experiments (Brems et al., 1981). The many photoactivatable derivatives of the cardiac glycosides and their aglycones exemplify the use of non-competitive inhibitors for affinity labeling (see Hall and Ruoho, 1980; and references therein). The inhibition of Na,K-ATPase activity was used to characterize these reagents. Molecules that act as allosteric effectors may also be characterized by enzymatic assays. Pfeuffer (1977) showed that P^3-(4-azidoanilino)-P^1-5′-GTP activates membrane bound adenylate cyclase, and Brunswick and Cooperman (1971) showed that the $O^{2'}$-ethyldiazomalonyl derivative of cAMP reverses the substrate inhibition of phosphofructokinase, by ATP, albeit less effectively than cAMP. Active derivatives of coenzymes have also been tested (e.g. for coenzyme A see Lau et al., 1977ab), as have competitive inhibitors of coenzymes (e.g. NAD: Standring and Knowles, 1980; and references therein).

Biochemical assays for measuring the potency of a photolabile ligand may not always be feasible. Often a direct or indirect binding measurement is the method of choice. Of course when the photoaffinity reagent has been made in labeled form such measurements can easily be conducted. But it is often more convenient to measure the competition of unlabeled reagent for the binding of a radiolabeled ligand that is more readily obtained. In this way several unlabeled derivatives can be tested using a single radioactive reagent; for example, Yip et al. (1980) used monoiodinated insulin in experiments to measure the affinity of arylazido insulins for the insulin receptor of rat liver plasma membranes, and several other iodinated peptides have been used in the same way (Carney et al., 1979; Lee et al., 1979; Niedel et al., 1980). Methods that have been used to measure binding constants for photoaffinity labels include equilibrium dialysis (e.g. a diazirino NAD analog to lactate dehydrogenase: Standring and Knowles, 1980), gel filtration (for photolabile haptens to antibodies: e.g. Smith and Knowles, 1974) and the recovery of membrane bound ligands by centrifugation (e.g. insulin derivatives: Yip et al. 1980).

Because they are usually chemically inert (the first use of an arylazide reagent was to stimulate the production of antibodies against it in rabbits (Fleet et al., 1969)), photoaffinity reagents may be used in biological assays in vivo or on tissue preparations, as well as in vitro. Examples of such assays include the demonstrations that 8-azido-cAMP is a weak chemoattractant for *Dictyostelium discoideum* (Wallace and Frazier, 1979a) and that members of a group of angiotensin II analogs were able to induce the contraction of rabbit aorta strips (Escher et al., 1978).

The investigator should be aware of a few possible pitfalls in binding studies and other assays. First, the reagent should be pure. This is of the utmost importance if the photoaffinity label is a derivative of the natural ligand and turns out to be less potent in its biological activity. In numerous cases where a derivative has appeared to be (say one hundred times) less active than its precursor, it has not been shown that the precursor was not a contaminant (say 1%) of the derivative which was actually completely inactive. This problem might be resolved by adding increasing amounts of *receptor* to samples of *ligand* of a fixed concentration. If a plateau of activity or binding is reached, it will reveal the concentration of active species in the ligand preparation (for example see Richards and Vithayathil, 1960).

The stability of the reagent to the assay conditions should also be ascertained. The instability of azides and diazo compounds towards thiols was mentioned earlier, and it would be prudent to ensure that all new reagents whether they contain familiar photoactivatable groups or not are stable to the prevailing chemical and enzymatic conditions. In an attempt to label the cAMP receptor of *D. discoideum*, Wallace and Frazier (1979b) found that 8-N_3-cAMP was converted to N_3-AMP by the phosphodiesterase of a crude membrane preparation. The N_3-AMP specifically labeled actin and not the cAMP receptor.

Last, the investigator should take care not to assume that an otherwise thoroughly investigated reagent will work in his system. Again instructive examples may be found amongst the various 8-azidoadenine nucleotides, some of which are commercially available as ^{32}P-labeled reagents. Substitution of the 8-position shifts the syn-anti conformational equilibrium which obtains in these molecules toward the syn form. While 8-azidoadenine nucleotides have proved useful in numerous photoaffinity labeling experiments there are also many examples in which binding to the desired site could not be demonstrated. Occasionally this may prove misleading; it seems that 8-N_3-cAMP labels phosphofructokinase at the catalytic site and not at the allosteric sites as expected (Lascu et al., 1979; Gottschalk and Kemp, 1981). The reader should not despair, as in other cases the specificity of the binding site is much weaker. Recently 2-nitro-4-azidophenylphosphate (Lauquin et al., 1 980) was used successfully as an analog of inorganic phosphate!

4.2. Sample preparation for photolysis

The first step in an actual photoaffinity labeling experiment is to allow the ligand to bind to its receptor. In most cases the experimenter wishes to obtain an easily measureable amount of site-specific labeling and at the same time keep labeling outside the receptor site (non-specific labeling) at a negligible or low level. To do this the ratio of bound ligand ($[L]_B$) to the total amount of ligand added ($[L]_{B+F}$) is set as close to unity as is reasonable. For a simple binary system:

$$R + L \rightleftharpoons R \cdot L$$

... the ratio $[L]_B/[L\text{-}]_{B+F}$ is given by:

$$\frac{([L]_{B+F} + [R]_{Total} + K_d) - \sqrt{([L]_{B+F} + [R]_{Total} + K_d)^2 - 4\,[R]_{Total} \cdot [L]_{B+F}}}{2[L]_{B+F}}$$

In other words, the receptor concentration, $[R]$, should be as high as possible and to achieve this it may be useful to carry out a partial purification of the receptor. The ligand, L, on the other hand must be kept at a low concentration, but one commensurate with obtaining a useful extent of photoaffinity labeling. This is illustrated in Table 4.1. Receptor–ligand complexes with low dissociation constants (< 1 μM) present no special problems but in cases where K_d is high it may be necessary to use high concentrations of ligand to occupy enough receptor sites for there to be a reasonable chance of obtaining a measurable extent of labeling. The free ligand in such cases can be a source of severe non-specific labeling and may further necessitate prolonged irradiation because of a shielding effect (see below).

Not all ligands bind rapidly to their receptors and sufficient time should be allowed for equilibration before photolysis is initiated. If the appropriate controls were done, slow binding will have been detected and quantitated when the K_d of the ligand was determined. With certain tight-binding ligands the rate of dissociation from the receptor may be slow enough that excess reagent can be removed before the irradiation step. This has been used to advantage in several labeling experiments to reduce non-specific labeling. Fleet et al. (1972) separated excess photolabile hapten from a hapten–IgG complex by gel filtration. In studies of the estrogen receptor

TABLE 4.1
The reversible binding of a ligand to a receptor: a hypothetical case ($K_d = 1\,\delta$ M)

Receptor concentration (μM)	Total ligand concentration (δ M)	% Total ligand bound to receptor	% Receptor sites occupied
0.1	0.01	9.0	0.9
0.1	0.1	8.4	8.4
0.1	1	4.9	48.8
0.1	10	0.9	90.8
0.1	100	0.1	99.0
1.0	0.01	49.9	0.5
1.0	0.1	48.8	4.9
1.0	1	38.2	38.2
1.0	10	9.0	90.1
1.0	100	1.0	99.0
10.0	0.01	90.9	0.1
10.0	0.1	90.8	0.9
10.0	1	90.1	9.0
10.0	10	73.0	73.0
10.0	100	9.9	98.9

Katzenellenbogen and coworkers (e.g. Katzenellenbogen et al., 1974) were able to exploit the very slow dissociation rate of the ligands. In such cases it may be possible to selectively destroy unbound reagent as suggested by Staros et al. (1978). It has even proved possible to administer ligands to live animals and allow them to bind to receptors in vivo. The approproate tissue is subsequently dissected and subjected to irradiation in vitro. Examples include the benzodiazepine receptor of rat brain, labeled with flunitrazepam (Mohler et al., 1980), and the acetylcholine receptor of rat diaphragm, labeled with an α-bungarotoxin derivative (Nathanson and Hall, 1980).

Occasionally, the poor solubility of the reagent in aqueous solution is a problem. It usually proves possible to circumvent this difficulty by adding the ligand to the solution of receptor in a small amount of an organic solvent that is miscible with water such as ethanol, dioxane, dimethylsulfoxide or dimethylformamide. Obviously solvents that absorb in the ultraviolet should be avoided unless short wavelength radiation is rigorously excluded, especially those such as acetone that might act as photosensitizers. The possibility that the addition of a solvent induces subtle alterations to the system under study should be investigated, perhaps by checking if the labeling results depend on the solvent concentration, within reasonable limits.

Finally, we must consider whether the sample should be photolysed under an inert atmosphere. Photooxidative damage to biological samples can occur, at short wavelengths, because of the direct reaction of triplet excited states of chromophores (e.g. tryptophan) in the sample with ground state (triplet) oxygen, or because of the generation of singlet oxygen by sensitizers in the sample. The sensitizers might be endogenous or the photoaffinity reagent itself might sensitize. Sensitizers can even present problems at wavelengths in the visible region. Deleterious effects of photooxidation include the loss of binding capacity of receptors for ligands, and the crosslinking of proteins. For example, Bayley and Knowles (1980) found that the proteins of red cell membranes became rapidly crosslinked when irradiated at 254 nm (but not at 300 or 350 nm), unless the preparation was purged with nitrogen.

Oxygen reacts with carbenes and nitrenes (e.g. Nielsen and Buchardt, 1982) and it is conceivable that it is advantageous to allow this reaction to take place, if damage to the sample does not occur. Presuming that no oxygen is present at the binding site of the receptor, the dissolved gas could act as a useful scavenger of unbound photoaffinity reagent, partly preventing non-specific labeling (see Section 4.7.4).

To free a sample from dissolved O_2 it is usually sufficient to pass a stream of water-saturated nitrogen or argon over (not through, which may cause denaturation) the sealed, stirred solution of the ligand–receptor complex for about half an hour ($\sim$ 1 ml s^{-1} for a 2 to 3 ml sample). Purging may be continued during irradiation or the vessel may be sealed under a slight positive pressure (for details see Section 4.3). As an added precaution, azide ion (1 mM) has been used to quench singlet oxygen (Smith and Benisek, 1980), and carotenoids may protect membranes from photochemical damage.

It should be added that numerous photoaffinity labeling experiments have been performed, with success, in the presence of oxygen. (Some less commonly used reagents may even require the presence of oxygen for covalent attachment, Section 2.3.8.) Nevertheless, the prevailing atmosphere is an experimental variable that should be kept in mind, especially when working at short wavelengths. Last, but not least, it should be apparent that the removal of oxygen will not protect against radiation damage by prolonged irradiation, if attempts are made to photolyse reagents with low extinction coefficients or quantum yields at < 300 nm (Table 3.1).

4.3. Photolysis of the receptor–ligand complex

First in this section the apparatus required for photolysis is discussed, and second the determination of the optimal duration of irradiation.

The apparatus can be considered as having three components: a light source, a filter system, and the vessel containing the sample. In almost all experiments involving photogenerated reagents it is advantageous to use a simple experimental set-up and avoid elaborate and expensive equipment. The book by Calvert and Pitts (1966) remains a most valuable source of practical information on photochemical equipment. *The Chemist's Companion* (Gordon and Ford, 1972) has a useful chapter, and the book by Murov (1973) is an excellent compilation on information of all aspects of experimental photochemistry. Many of the catalogs of the suppliers of photochemical equipment contain detailed specifications of lamps and filters (see List of Suppliers).

For irradiation at 254 nm a low pressure mercury vapor arc lamp is used. The 185 nm light emitted by these lamps is absorbed by the lamp envelope and the vessel containing the sample. Ninety percent of the remaining light is at 254 nm and, if desired, the other weak spectral lines in the UV and visible can be removed with the appropriate filter (see below). As the pressure of vapor inside a mercury lamp is increased the 254 nm line becomes weaker relative to other spectral lines because of self-absorption. Useful emission in the 300 to 400 nm range is obtained from medium pressure mercury lamps which are usually considerably more intense (say 100 times per unit length) than low pressure lamps, and may require external cooling. The spectrum from a medium pressure arc contains several intense bands in the UV and visible (strongest at 365, 436, 546 and 578 nm). High pressure arcs (> 100 atm.) emit an almost continuous spectrum from 220 to 1,400 nm. They are unnecessarily intense for most purposes, require a rapid flow of water for cooling, and may be somewhat erratic in output. If high intensity radiation is required a useful alternative is a medium pressure point source such as a Xe, Hg or Xe-Hg short arc. Such lamps require no cooling, or just a fan, and the intense output may be readily collimated. Point sources usually operate at 20 to 100 atm. The mercury vapor sources emit several pressure broadened peaks corresponding to those of the medium pressure arcs, superimposed on a continuum. The xenon arcs emit a continous spectrum in the UV and visible with some sharp peaks in the near IR. Xe-Hg arcs have the charactsristics of mercury arcs but they are more stable and longer-lived. While medium pressure arcs and point sources are useful in the visible region, a simple alternative for work above 400 nm is to use an incandescent source such as a quartz tungsten-halogen filament lamp.

A few words on safety are necessary. Ultraviolet absorbing goggles, which are available from all major laboratory supply companies, should be worn when working with all but the lowest power short wavelength lamps. Skin exposure to UV irradiation should be minimized, and gloves or a sun-screen cream may be useful for the hands although there is no reason why levels of exposure requiring this should be reached. Light below ~ 200 nm produces ozone from atmospheric oxygen and high power (> 200 W) xenon or mercury point sources should be operated in a hood or preferably, to spare the optics, vented directly from the lamp housing. 'Ozone-free' lamps have an envelope that absorbs the offending radiation. High pressure lamps are an explosion hazard and must be handled with caution. Those containing xenon are pressurized even when cold.

An inexpensive commercial apparatus or equipment adapted from other uses usually makes a light source suitable for most experiments with photochemical reagents. I have found the Rayonet Model RMR-400 'mini-reactor' to be very useful. The apparatus comprises a single lamp (26 cm long) which is air cooled by a fan in the base of the reactor. The sample is usually clamped 3 to 4 cm from the center of the lamp and at its midpoint. Three interchangeable low power lamps are used: the RPR-2537 Å (a low pressure Hg vapor lamp with > 80% emission at 254 nm), the RPR-3,000 Å (emits predominantly 275 to 340 nm, with a band at 254 nm and bands in the visible), and the RPR-3500 Å (310 to 400 nm). The last two are phosphor coated low pressure Hg lamps. For convenience, the manufacturer's protective shield can be removed from the apparatus and the reactor placed inside a home-made enclosure with a thick black cloth curtain at the front. This will facilitate the positioning of a magnetic stirrer and lines carrying coolant to and from the reaction vessel. We have also used a fluorescence microscope illuminator containing a 200 W Hg point source, equipped with the original collimator and glass filters. Other workers have used ultraviolet lamps of the sort used to visualize fluorescence-quenching molecules on TLC plates, and they are surprisingly powerful, photolysing dilute solutions of aryl azides in a few minutes at 1 to 2 cm. Models such as the Mineralight UVSL-25 have two lamps emitting predominantly at 254 nm and at 366 nm. To irradiate in the visible region (e.g. for nitroazidoanilines) a slide projector, which uses a powerful (e.g. 250 W) tungsten-halogen lamp can be used. The more affluent investigator can choose from the great variety of lamps sold by the companies listed in the Appendix.

To remove unwanted wavelengths several classes of filter may be used including chemical filters (solutions of light absorbing substances), glass filters and interference filters. Sophisticated monochromators are not normally useful as the narrow bandwidth that they provide is rarely necessary and in producing it they reduce the usable light intensity enormously, hence a powerful light source is required to achieve reasonably short photolysis times.

Low band-pass interference filters (10 nm width) are available throughout the UV and visible regions with peak transmissions ranging from 10 to 50%. Inexpensive broad band-pass filters (70 nm bandwidth) are more useful for the present applications and have peak transmissions of > 60%. Unfortunately such filters are not available in the UV region. Numerous glass filters are available that absorb throughout the UV-visible spectrum and by using them it is possible to isolate almost any broad region of the spectrum and completely cut off unwanted short wavelengths. Pyrex cuts out light of < 290 nm and may be used to remove the residual 254 nm output from the RPR-3000 Å and RPR-3500 Å lamps described above. Neutral density filters are useful for attenuating intense sources.

Chemical filters have been described in detail by Calvert and Pitts (1966). They are solutions contained in optical cells of appropriate path-lengths. For the purpose of photoaffinity labeling it is usually required to limit the light flux at the damaging short-wavelength end of the spectrum and the elaborate two or three solution filters described by Calvert and Pitts are not necessary if some light from the visible and IR can be tolerated. For example, glacial acetic acid (1 mm pathlength) has been used to filter light below 230 nm. Light below 310 nm can be removed with filters containing saturated copper sulfate (useful up to 550 nm) or 2% (w/v) potassium hydrogen phthalate. Cobaltous nitrate (40% w/v) provides a useful window (e.g. for photolysing diazirines) in the near UV at approximately 340 to 380 nm (Isaacs et al., 1977). Aqueous sodium nitrite (1.0 M) absorbs radiation below 380 nm, and is useful for filtering tungsten-halogen lamps for irradiations in the visible region. Powerful arc lamps and quartz halogen lamps produce much infrared radiation. Heating of the sample can be prevented by using a thick pyrex filter or a liquid filter of 10 cm of water. A heat reflecting mirror (hot mirror) might also be used but those available are not transparent to ultraviolet radiation. Filters can and should be checked spectrophotometrically.

It is emphasized that for most purposes it is satisfactory to carry out irradiations with low intensity lamps without filters other than the sample containers. Irradiations, with the RPR-2537 Å or equivalent are done using quartz vessels (below) and with the RPR-3000 Å or 3500 Å using glass vessels.

For samples of 50 μl to several milliliters quartz or glass test tubes (or optical cuvets) are satisfactory vessels for irradiating samples. For example, a sample of 1 ml may be sealed with a rubber serum cap inside a 13×100 mm test tube containing a magnetic stir-bar. Wet N_2 gas is passed slowly over the stirred sample using syringe needles to introduce and exhaust the gas. To seal the vessel under a slight positive pressure the exit needle is removed first. If irradiation must be prolonged (> 15 min) the N_2

flow or a slight positive pressure of the gas may be maintained during irradiation. More elaborate vessels with an outer jacket through which water, or a chemical filter, can be pumped to maintain a constant temperature (e.g. 37 °C) may also be constructed from quartz or glass (Fig. 4.1). When the sample scatters light strongly (e.g. membranes) or is otherwise opaque, constant stirring during irradiation is essential.

When many samples are irradiated at once care should be taken to clamp them in geometrically equivalent positions around the lamp which will be, preferably, a long vertical source, and not a point source. Alternatively, a rotating tube holder or 'merry-go-round' may be used to ensure even exposure to the radiation. (Such an accessory, RMA-400, is available for the Rayonet mini-reactor.) Many small samples may be photolysed together by placing them in wells in a microtiter dish. A Mineralight TLC plate visualizer is a convenient source in this case but care must be taken to ensure that the samples are equally illuminated. The samples may be agitated magnetically using fleas made from pieces of paper clip sealed in glass Pasteur pipette tips and if necessary an inert atmosphere may be maintained by placing the sample holder and lamp in a small plastic bag and passing nitrogen or argon through it. Very small samples (< 5 μl) can be sealed in glass or quartz capillaries, and samples of high absorbance may be irradiated as thin films, e.g. on microscope slides for small volumes, or

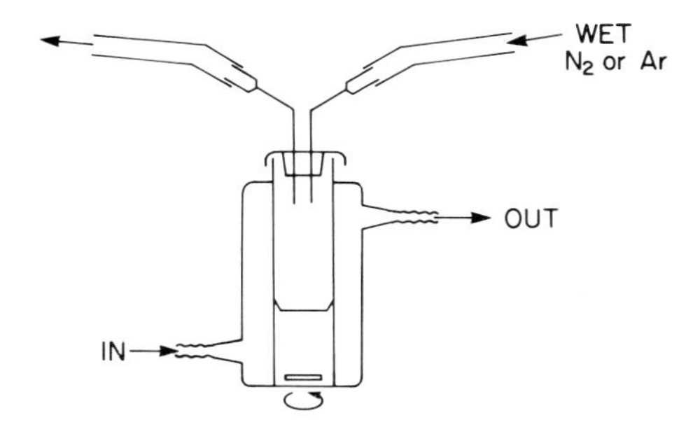

Fig. 4.1. Photolysis under an inert atmosphere. The sample is contained in a quartz or glass vessel and stirred with a magnetic stir-bar. The vessel is jacketed and kept at constant temperature with a circulating fluid that can also act as a filter. An inert atmosphere is maintained by passing wet argon or nitrogen (the gas from a tank is bubbled through water) over the solution at ~ 1 ml s^{-1}.

spread out on the internal surface of a horizontal rotating tube for larger volumes (Staros and Richards, 1974).

The duration of photolysis is important. Ideally the extent of covalent attachment of the ligand to the receptor should reach a plateau after adequate irradiation. The extent of labeling is determined by using a radiolabeled ligand (Section 4.6) or by using a photoinactivation assay as described in the next section. Occasionally the extent of covalent attachment will decrease if irradiation is prolonged. This occurs if the bond to the receptor is photolabile or if the radiolabel itself is photolabile. Alternatively, slow incorporation of the ligand may continue to occur after the rapid initial phase. This occurs if the products of photolysis of the reagent can themselves label the receptor albeit less efficiently than the ligand designed to do the job. (If a photoinactivation assay is used there are several other possible explanations for analogous effects; see Section 4.4.)

In most cases the investigator will attempt to maximize receptor-specific labeling by determining the appropriate duration of photolysis under defined conditions. If either the receptor preparation or radiolabeled ligand is a scarce commodity an approximation of the photolysis time required may be obtained by irradiating a sample of unlabeled photoaffinity reagent in buffer or solvent and determining the rate of photolysis by measuring the ultraviolet or visible absorption spectrum at appropriate intervals. At low concentrations the rate of change of concentration of a photosensitive component under irradiation is given by:

$$-\frac{\mathrm{d}a}{\mathrm{d}t} = \frac{\psi\,(1 - \mathrm{e}^{-2.303\,\mathrm{o}\,a\,l})\,I_0}{l}$$

$$\simeq 2.303\,\psi\,I_o\,\varepsilon\,a$$

where: a = molar concentration of the photolabile species
t = time
ε = molar extinction coefficient
l = length cuvet (cm)
I_0 = light intensity (mE s^{-1} cm^{-2})
ψ = quantum yield of photolysis

So, when light absorption by the sample is low, the fraction of photoaffinity reagent (F_T) photolysed in a time interval (T) is independent of its concentration, and of the pathlength of the cell:

$$F_T \simeq 1 - e^{-2.303\, \varphi\, I_0\, \varepsilon\, T}$$

$$T_{1/2} \simeq \frac{0.3}{\varphi\, I_0\, \varepsilon}$$

Therefore the $T_{1/2}$ for photolysis, which is independent of concentration, determined for an unlabeled reagent, or a model compound with similar structural features, should approximate that in the receptor–ligand complex, as long as the light absorption and scattering of the sample is low at the wavelengths in question. The value of such a determination is limited, however, because the quantum yield for photolysis of the reagent may be different at the receptor binding site if alternative pathways for quenching of the excited state are available or those that occur in solution are slowed.*

A photolysis time of 5 to 10 min should be aimed for and may first be established with the model compound by varying the distance of the sample container from the lamp (for an uncollimated source) or by attenuating the source with neutral density filters (for a collimated beam). This duration is conveniently short and yet not so brief that split-second timing is required for reproducible results.

Having obtained a rough estimate of the irradiation time, the time dependence of labeling of the receptor should be measured directly. The optimal photolysis time (e.g. Fig. 4.2), determined by the incorporation of label or by photoinactivation (see below), will be used in many labeling and control experiments, and it is important that it be reproducible. For the results to be useful the sample must be irradiated at a fixed point relative to the lamp. For example, if a long arc is used, the intensity varies both with the distance from the center of the lamp *and* with the position along the

* Extraordinary differences between photolysis in buffer and in biological samples have occasionally been noted (e.g. Hosang et al., 1981). An increased rate of photolysis due to the environment of the active site is advantageous (Section 2.4).

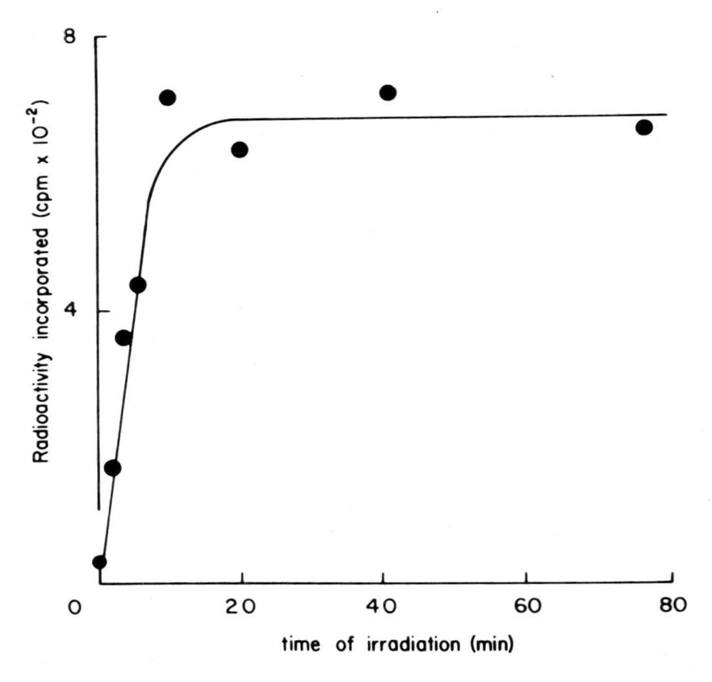

Fig. 4.2. Covalent incorporation of [^{3}H]3-(3H-diazirino)pyridine adenine dinucleotide into lactate dehydrogenase as a function of time. Lactate dehydrogenase (1.1 mM in subunits) in 100 mM Tris–HCl at pH 8.0 was mixed with [^{3}H]3-(3-H-diazirino) pyridine adenine dinucleotide (1.2 mM) at 4°C. The mixture was irradiated with the RPR 3,500 Å lamp of a Rayonet minireactor and at the times indicated portions were analyzed for the incorporation of radiolabel. (From Standring and Knowles, 1980.)

length of the lamp. Obviously all components of the system (the vessels, filters, temperature, etc.) should be the same in all experiments. A most important variable is the output of the lamp which will vary with its temperature, the warm-up time, and its age. For consistent results the lamp should be switched on an hour before an experiment and samples should be removed without turning the lamp on and off. The warm-up and stability of medium pressure arc lamps and point sources can be monitored with the voltmeter and ammeter of the power-supply. New lamps often change their characteristics in the first few hours of operation and they should be run for three or four hours before use in an experiment.

The output of a lamp may be monitored by chemical actinometry. The standard method remains the irradiation of potassium ferrioxalate which is useful in the range 254 to 480 nm. The method is described in detail by Calvert and Pitts (1966) and by Murov (1973). But perhaps, the most convenient way to check the lamp output is to keep a stock solution of a model compound or the reagent itself, if it is readily available, and determine its rate of photolysis periodically by irradiation in a spectrophotometer cuvet. In this way the output of the lamp in the region of interest can be rapidly checked. An alternative is to use one of many radiation measuring instruments that are commercially available, a thermopile and voltmeter, for example (see Appendix for a list of manufacturers).

4.4. Photoinactivation experiments

Photoinactivation (or occasionally photoactivation) experiments have often been used to test the utility of photogenerated reagents. The receptor preparation is incubated with an appropriate concentration of a photoaffinity label and then irradiated (see Section 4.3 for details). After dilution, or removal of excess ligand and photolysis products by dialysis or gel filtration, the preparation is assayed for activity or ligand binding. If an enzyme active site has been blocked in the experiment, the enzyme will no longer be active; if the receptor for a toxin has been blocked it will no longer bind toxin.

Using photolabile derivatives of tetrodotoxin Chicheportiche et al. (1979) produced irreversible inactivation of sodium channels in isolated nerves as determined by an electrophysiological assay. Wrenn and Homcy (1980) induced irreversible blockage of β-adrenergic receptors, and Staros and Knowles (1978) found inhibition of dipeptide transport in *E. coli* after irradiation of living cells in the presence of a dipeptide photoaffinity reagent. After several cycles of irradiation in the presence of a cholecystokinin analog, pancreatic ascinar cells began irreversible secretion (Galardy et al., 1980).

Such an approach to the preliminary evaluation of photoaffinity reagents goes one step beyond the reversible binding experiments discussed earlier. It is a useful approach in that, like competitive binding experiments, radiolabeled ligand is not required and therefore many potential reagents may be screened. However, many pitfalls severely reduce its utility.

In many cases when the number of receptors blocked in a photoaffinity labeling experiment has been estimated using a radioactive reagent (see Section 4.5), and activity measurements have been made on the same samples there has been no relationship between the number of sites specifically labeled as judged by the two criteria. In other cases a striking parallel has been found. Chen and Guillory (1981) obtained excellent photoinactivation data when an arylazide derivative of NAD was irradiated with mitochondrial NADH dehydrogenase. Kaczorowski et al. (1980) found that irreversible loss of activity of the lactose transporter of *E. coli* corresponded well with the incorporation of label from 4-nitrophenyl-σ-D-galactopyranoside. By extrapolating the linear relationship between the inactivation of acyl-coenzyme A: glycine *N*-acyltransferase and the incorporation of low levels of *p*-azidobenzoyl-CoA, Lau et al. (1977) found a one-to-one relationship between inactivation and covalent attachment (see Fig. 4.3). In contrast, Carlier et al. (1979), for example, found no correlation between the extent of labeling with ADP and ATP analogs and inactivation of spinach CF_1. Fannin et al. (1981) found that photoinactivation of mutarotase with [^{3}H]phloretinyl-3′-benzylazide was 5 to 6 times greater than label incorporation. Payne et al. (1981) found a correlation between the covalent attachment of 4-azido-[^{3}H]estradiol to rat α-fetoprotein and the loss of estradiol binding capacity. On careful investigation, however, it was shown that the correlation was fortuitous and that little of the attached label was at the steroid binding site.

The not infrequent failure of the extent of attachment of photoaffinity labels to parallel receptor inactivation can be explained in several ways. First, in some membrane-bound receptor systems, receptor subunits are present in excess over the other components of the system. Second, the ligand might induce or take part in a photochemical reaction at the ligand binding site that does not culminate in covalent attachment. Ligand sensitized photooxidation can be prevented by irradiating the sample in the

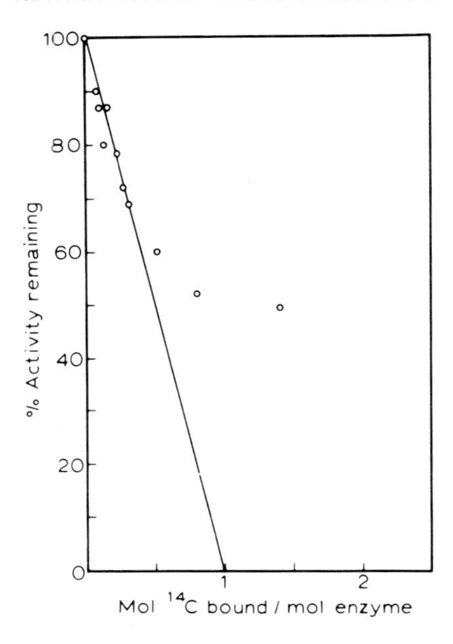

Fig. 4.3. Photoinactivation of acyl-CoA : glycine *N*-acyltransferase by *p*-azido[^{14}C]benzoyl-CoA. At low extents of labeling the fraction of enzyme inactivated was directly proportional to the amount of the photoaffinity reagent that became attached. (From Lau et al., 1977a.)

absence of oxygen but little can be done to prevent or easily detect more subtle changes. Benisek and coworkers have given detailed accounts of the active site modification of ketosteroid isomerases by α,β-unsaturated steroid ketones. In two separate cases (Ogez et al., 1977; Smith and Benisek 1980) inactivation of the enzymes occured without coupling of the reagent to the protein. A single amino acid residue in the active site was found to be modified in each case. It is instructive to note that in early experiments tight binding of the photolysed ligand was mistaken for covalent attachment (compare Martyr and Benisek, 1973, 1975 and Ogez et al., 1977). Third, the tight binding of photolysis products can itself lead to discrepancies between values for apparent covalent binding and inactivation. It has been noted on several occasions that photodecomposition products derived from the reagent may bind far more tightly than the ligand itself, and may only be removed under denaturing conditions (e.g. Fisher and Press, 1974; Standring and Knowles, 1980). In yet other cases, quite unexpected results have been obtained. When an active analog of angiotensin II was attached to its receptor, the receptor was not permanently activated, but irreversibly inactivated (Escher et al., 1978).

Several controls related to those required in photoaffinity labeling experiments with radiolabeled ligands (see Section 4.7) must also be performed in studies of photoinactivation. In particular inactivation should not occur on irradiation in the absence of the photoaffinity label, and the receptor site should be protected against photoinactivation if it is first blocked with a photochemically inert molecule.

4.5. Experiments with radioactive reagents

In discussing photoaffinity labeling experiments with radioactive reagents, I shall focus mainly on the labeling of protein receptors with small photolabile ligands. The same principles may be applied to related experiments. The general approach to identifying a receptor in a mixture of proteins is first described. The 'mixture' may be extremely complex, e.g. whole cells or a membrane preparation, or relatively simple, e.g. a purified multisubunit protein. Second, experiments to locate a label within a protein are described. This may be at the level of a low resolution peptide map or in favorable cases the identification of a labeled amino acid residue.

4.5.1. Identification of labeled receptor subunits

The method of choice for identifying a labeled receptor in a mixture of polypeptides is SDS-polyacrylamide gel electrophoresis. 'Identification' usually means the assignment of a rough molecular weight to the labeled polypeptide as judged by its relative rate of migration in the gel. The labeled protein band is detected by autoradiography, fluorography or by cutting and counting the gel. Successful examples, chosen from many, include the labeling of cAMP binding proteins of sarcoma cells (Skare et al., 1977), the labeling of the benzodiazepine receptor in rat brain (Fig.

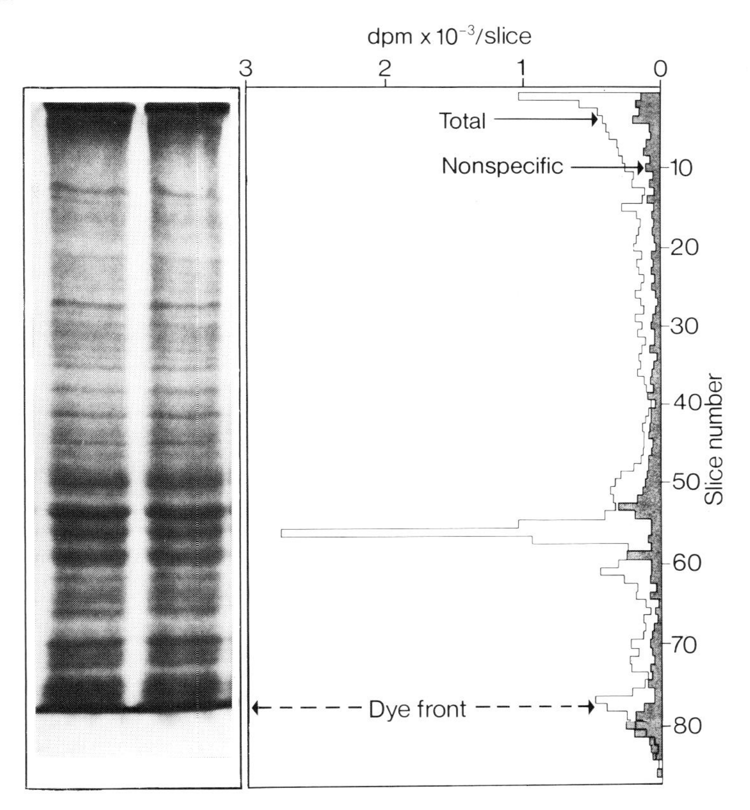

Fig. 4.4. Photoaffinity labeling of a benzodiazepine receptor. SDS-polyacrylamide gel electrophoresis of purified synaptic membranes after photoaffinity labeling with [^{3}H]flunitrazepam (3 nM). *Right:* distribution of radioactivity in the gel. The hatched area is the label distribution when diazepam (10 μM) was present during photolysis (non-specific labeling). *Left:* Coomassie blue staining pattern after irradiation (with and without diazepam present). For details see Mohler et al. (1980).

4.4, Mohler et al., 1980), and the identification of the insulin receptor (Section 1.2.1). In all these examples specific labeling was achieved in extremely complex mixtures of polypeptides.

Unbound radioactive photolysis products usually run close to the dye-front on electrophoresis. However, exceptions have been noted in which small molecules run in other regions of gels and appear as quite sharp bands (for example see Thomas and Tallman, 1981). Macromolecular reagents may become crosslinked on photolysis, especially if they are multisubunit proteins. The species formed will be of lower electrophoretic mobility and will usually correspond to those formed on irradiating the reagent alone (e.g. Ji and Ji, 1981). It is advisable then to remove non-covalently bound material by dialysis, gel filtration or, for membranes, centrifugation and resuspension before electrophoretic analysis. Photolysis products may bind more tightly than the photoaffinity reagent to the receptor and there may be no convenient way to remove them without disrupting the receptor. As there is no need to isolate the labeled receptor in the native form for SDS-polyacrylamide gel electrophoresis, the precipitation of proteins with trichloracetic acid or acetone may be a useful way to remove non-covalently bound ligands of low molecular weight. A useful control, to confirm that free derivatives of the ligand in the gel, is to treat a portion of the labeled preparation with a proteolytic enzyme before electrophoresis and check that the labeled band disappears or at least increases in mobility. Nicolson et al. (1982) found that non-covalently bound photolysis products of *p*-azidopuromycin stuck tightly to ribosomal proteins and ran with them in urea-polyacrylamide gels. A thorough washing removed this material from the ribosomes. The proteolysis experiment suggested above would have been misleading in this case, but the radioactive contaminants might have been removed had an SDS-polyacrylamide gel been used. An extensive discussion of control experiments that ensure the validity of assigning a radioactive band in an SDS-polyacrylamide gel to a receptor polypeptide is given later (Section 4.7).

Although many workers have not proceeded further than assigning a molecular weight to a labeled receptor the method should in principal be useful for following the purification of receptors (e.g. Newman et al. 1981; Shorr et al. 1982). However, great care must be taken as the covalently attached reagent can affect the physical properties of the receptor and there is no reason to expect that the unlabeled and labeled species should copurify under all conditions. A more extensive discussion of a related problem is given below when the purification of labeled peptides generated by cleavage of a receptor polypeptide is considered. The caveat extends even to SDS-polyacrylamide gel electrophoresis which is supposed to separate according to molecular weight. Of course when the photoaffinity reagent is large, such as a modified insulin molecule, its mass must be taken into account when the molecular weight of the labeled polypeptide is estimated. But even when the mass of the ligand is small problems arise. Most seriously, it is not always possible to assume that a Coomassie blue stained band that migrates at the same position as the label is the stained receptor. Even in an SDS-gel the mobility of the polypeptide might be slightly altered after reaction with a photoaffinity reagent, if a charged group is involved (see for example Noel et al., 1979). This is particularly troublesome in a complex system containing numerous polypeptides but usually of little consequence in a simple system with only a few well-separated components. For example, a photolabile herbicide, azido-atrazine, labeled a 32,000 dalton polypeptide in a chloroplast preparation that was not quite coincident with a major Coomassie blue stained band (Pfister et al., 1981). It seems likely, however, that the stained band represents the protein that binds the herbicide as it is absent in certain mutants that do not bind atrazine.

Two-dimensional gel systems that combine isoelectric focusing or electrophoresis in the absence of a charged detergent in the first dimension with electrophoresis in SDS in the second dimension present more serious difficulties as labeled and unlabeled species may separate widely in the first dimension. The resolving power of isoelectric focusing was dramatically demonstrated by the work of Swanson and Dus (1979) who labeled a purified cytochrome *P*-450 with *N*-(azidophenyl)imidazole. The protein contained two binding sites for the reagent and the unlabeled, once labeled and twice labeled polypeptide chains were easily separated by IEF. It can only be concluded that such a result means that powerful two-dimensional separation systems such as that of O’Farrell (1975) in which several thousand polypeptides can be separated are of limited use for correlating stained bands and labeled bands in photoaffinity labeling experiments. A further complication is that a polypeptide labeled at a single site may separate into several radioactive bands on isoelectric focusing if the polypeptide is heterogeneous in charge because a variety of functional groups at the active site have reacted with the reagent. A case in point is the identification of cAMP binding proteins of red cell membranes with $[^{32}P]8\text{-}N_3\text{-}cAMP$ (Hoyer et al., 1980). Labeling products were obtained which appeared heterogeneous in the IEF dimension on two-dimensional gels but most likely arose from single labeled receptor polypeptides. Ribosomes are an excellent model system for demonstrating such phenomena as the polypeptide subunits have all been identified. When the polypeptide L4 was crosslinked photochemically to the 23 S RNA of *E. coli* ribosomes and the adduct exhaustively digested with ribonucleases the resulting L4 derivative did not comigrate with L4 in a two-dimensional gel system (Maly et al., 1980). Cooperman et al. (1975) also found that photoaffinity labeled ribosomal subunits had somewhat altered electrophoretic mobilities.

In certain special cases the identity of a labeled band may be confirmed independently. Grant et al. (1979) used antibodies against identified ribosomal subunits to immunoprecipitate labeled polypeptides. When Pfeuffer (1977) attempted to label a GTP-binding membrane component that controls adenylate cyclase activity he found four labeled bands on SDS-polyacrylamide gels. Only one of these was associated with adenylate cyclase activity on sucrose density gradients. Girshovich et al. (1976) also using a GTP analog were able to show that only elongation factor G was labeled in a ribosome–EFG complex as the intact ribosome was easily separable from EFG by gel filtration.

Low molecular weight photoaffinity reagents rarely label more than one subunit of a single receptor, although this has been observed with cardiotonic steroid derivatives and the α- and β-subunits of Na,K-ATPase (Hall and Ruoho, 1980). With macromolecular reagents the specific labeling of several receptor subunits has been noted more often. One explanation for this is that macromolecular reagents are often mixtures of derivatives (Section 3.6.3) with photoactivatable groups attached at positions that may be some distance apart and lie adjacent to different regions of the receptor in the receptor–ligand complex. Second, the photoactivable group is usually on an arm capable of sweeping through a large volume in the region of the receptor, i.e. most of these molecules can be considered as exo-affinity reagents (Section 3.6.1; Fig. 3.26).

Ji and Ji (1981) labeled the choriogonadotropin (hCG) receptor with derivatives of hCG, a polypeptide hormone with two subunits (α: 14,000; and β: 23,000). The α-subunit was modified with an azidoaryl group, radiolabeled, and recombined with the β-subunit. On irradiation with a preparation containing the receptor, four polypeptides were labeled as judged by SDS-polyacrylamide gel electrophoresis. In a complementary experiment the β-subunit was similiarly modified and then recombined with α. Three polypeptides were labeled with this reagent, all of which corresponded in molecular weight to polypeptides labeled by the first derivative in which α was modified. Similar results were obtained whether the azido group was at the end of a 7 Å, 10 Å or 13 Å arm.

Bisson et al. (1980) labeled cytochrome oxidase with yeast cytochrome *c* to which an azido aryl group had been attached at Lys-13, and demonstrated that subunit II of the oxidase reacted specifically. In contrast, subunit III was attacked by a cytochrome *c* with an azido aryl group attached at Cys-102 (Moreland and Dockter, 1981). Lysine-13 and Cys-102 are on opposite sides of the cytochrome *c* molecule.

4.5.2. Identification of labeled peptides and amino acids

The usual techniques of protein chemistry have been used in attempts to identify the regions of polypeptide chains labeled by photoaffinity reagents. At low resolution there are usually few difficulties if peptides are separated on the basis of size: by gel filtration or SDS-gel electrophoresis. The identification of a 10,000 dalton peptide derived by cleavage from a purified 50,000 polypeptide will be no more difficult than the identification of a labeled polypeptide in a mixture of species. When the level of resolution is increased to the point where small peptides and modified amino acids must be identified severe problems arise. Unless the stoichiometry of labeling is close to 1 : 1, methods that separate on the basis of anything but size should be avoided as the labeled peptide may be separated from the unlabeled peptide. This will make the task of characterising the labeled species highly demanding because of the paucity of material. Further, it may be confusing if the labeled peptide comigrates or cochromatographs with a different, unlabeled species. Methods for increasing the stoichiometry of labeling and thus alleviating this problem will be discussed later (Section 4.6). Even if a high extent of labeling has been achieved the indiscriminate nature of photogenerated reagents may for once be a disadvantage. A single site may be derivatized at numerous positions which are close together in space. At worst several labeled peptides might be formed on cleavage of the protein each of which is modified in a number of different ways.

In practice, clear cut results have been achieved, including a few cases in which a single amino acid is modified, but more or less intractable situations have also arisen. Some illustrative examples are now described.

In a conspicuously successful series of experiments Kerlavage and Taylor (1980) labeled the regulatory subunit of cAMP-dependent protein kinase II with 8-N_3-cAMP. Up to 0.5 mol of reagent was incorporated per mol of polypeptide and excess reagent was removed by gel filtration on Sephadex G25 in the presence of 6 M guanidine·HCl. A tryptic map revealed a single labeled peptide and the site of attachment was investigated by CNBr cleavage of the carboxymethylated subunit. Gel filtration of the peptides on Sephadex G50 yielded a single peak of radioactivity containing a number of peptides which were further separated by HPLC on a reverse phase column. Two important observations were made: first, the labeled and unlabeled peptides were easily separated; second, each peak was a doublet comprising peptides containing the open and closed form of the carboxyl terminal homoserine lactone. Amino acid analysis and Edman degradation revealed that the Tyr in position of 7 of a 14-residue peptide had been modified. This was the only modification. Had the situation been more complex (i.e. if several labeled peptides, in consequently reduced yields, together with a larger amount of unlabeled peptide had been produced), analysis by a powerful separation method such as HPLC would have proved intractable.

In an early example, Fleet et al. (1972) also achieved a high stoichiometry of labeling when immunoglobulins were labeled with the hapten N^{α}-(4-azido-2-nitrophenyl)lysine. This aided them in tracking down a site of labeling in the heavy chain. Pure labeled peptides were obtained after (a) cyanogen bromide cleavage followed by gel filtration, (b) reduction, carboxymethylation, and trypsin digestion followed by gel filtration, and (c) paper electrophoresis of a peak from the latter. The last step yielded a tripeptide and a tetrapeptide and the sites of modification were identified as Lys-92 and Ala-93 on the basis of the absence of these residues on amino acid analysis. Only 13 % of the incorporated radioactivity was recovered in these two positions. The remainder could not be located but part of it was likely to be in the hypervariable region and would be extremely difficult to characterize.

In another successful case, Hexter and Westheimer (1971) were able to locate 5 % of the total radioactivity in Tyr-146 after irradiation of [^{14}C]diazoacetylchymotrypsin. The reaction is actually intermolecular, occurring in chymotrypsin dimers. Westheimer's group have determined the structure of several of the modified amino acids derived from the photolysis of proteolytic enzymes acylated with diazo reagents. Such data is not available for other photoaffinity reagents. Knowing that *O*-carboxymethyl tyrosine was an expected insertion product Hexter and Westheimer (1971) were able to show that of the two Tyr residues in the chymotrypsin B chain only Tyr-146, the C-terminal residue, was modified. If the nature of the modified amino acid had not been known it would have been considerably more difficult to pin-point the site of photolabeling.

Working with a low extent of labeling (2 %), Havron and Sperling (1977) were obliged to separate, by ion-exchange chromatography, unmodified ribonuclease from that which had been photocrosslinked to pUp. From the 1 : 1 adduct they were then able to produce a single labeled peptide by trypsin cleavage (Asn-67–Arg-85). Thermolysin cleavage of the tryptic peptide yielded two labeled peptides. In one both Ser-80 and Ile-81 were modified by a single ligand. In the other a derivative of Thr-82 was present.

Maly et al. (1980) irradiated *E. coli* 50 S ribosomes and found that protein L4 was specifically crosslinked to the 23 S RNA. Again a 1 : 1 adduct was isolated, and after pepsin digestion, label was found in two peptides. Amino acid analysis showed that the peptides were similar and that a Tyr residue (residue 45 of L4) was altered. In one peptide the altered residue was N-terminal and in the second peptide the residue was the fifth and modified differently from that in the first peptide, in such a way that Edman degradation was blocked.

It might be thought that Edman degradation would generally be useful for locating labeled amino acids but other groups have also experienced difficulties. For instance, Lifter et al. (1974) were able to assign two tyrosine residues labeled with 2,4-dinitrophenylazide in CNBr fragments of the heavy chain of an IgA molecule by automatic sequencing. The yields of released radioactivity were very low in both cases and the authors presented evidence that the Edman degradation might have failed at the modified residues. They also observed that CNBr cleavage was inhibited if a tyrosine adjacent to the reacting methionine was modified. Huang et al. (1982) were recently able to locate the site of labeling of bacteriorhodopsin with a retinal analog, *m*-diazirinophenylretinal (Fig. 3.12), by Edman degradation of a purified iodosobenzoic acid cleavage fragment. In this case approximately 25% of the radiolabel placed in the sequencer cup was recovered after 20 cycles, and about half of this was in a peak at cycles 4 and 5, corresponding to Ser-193 and Glu-194 of bacteriorhodopsin. The PTH-derivatives of modified amino acids often have solubility properties that differ from the standard PTH-amino acids and all extracts from the sequencer must be examined for radioactivity (Lifter et al., 1974; Bayley et al., 1981; Huang et al., 1982; Ross et al., 1982).

In the cases cited above labeled amino acids could be specifically identified. In many cases the investigator may only wish to know what region of the polypeptide chain the radioactivity is in. Further, it is often extremely difficult to extend the analysis beyond the peptide level because of the lack of specificity of photoaffinity labeling or because of technical problems. Brems et al. (1981), for instance, labeled the active site of prenyl transferase with *o*-azidophenethyl pyrophosphate. The extent of labeling was high if a repeated labeling procedure was used (see Section 4.6), and 80% of the incorporated label was recovered in a 30 residue CNBr fragment of the carboxymethylated polypeptide (86,000 daltons). In contrast to the cases cited above amino acid analysis, tryptic fingerprinting and Edman degradation revealed that label was spread throughout the peptide. The amino acid analysis was similar to the corresponding unlabeled peptide, suggesting widespread labeling, and a complete pronase digestion (to free amino acids) yielded eleven radioactive species. Tryptic digestion of the CNBr fragment gave at least three radioactive peptides, and on Edman degradation of the fragment, radioactivity was released in almost all the cycles. For the latter the pyrophosphate residue had to be removed by alkaline phosphatase treatment or else little radioactivity was recovered in the *n*-chlorobutane extract. The analysis could have been improved by demonstrating that the radioactivity eluted at each cycle was a PTH-amino acid, the most likely possibility, and not peptides washed out from the sequencer cup or label cleaved from the peptide by the harsh reagents used.

Some other cases in which problems were encountered in extending the resolution of the method to individual amino acids include the work of Bridges and Knowles (1974). Chymotrypsin was acylated with the *p*-nitrophenyl ester of *p*-azidocinnamic acid. After irradiation, the C-chain was the most heavily labeled but on further analysis many radioactive peptides were found. Fisher and Press (1974) experienced great difficulty in identifying N^{ε}-(4-azido-2-nitrophenyl)lysine labeled peptides from the heavy chain of an IgG preparation. Besides the heterogeneity of the polyclonal antibody, their problems included the loss of label from the peptides both on CNBr and trypsin cleavage and the generation of numerous labeled peptides. Aiba and Krakow (1980) labeled the cAMP binding protein of *E. coli* with 8-N_3-cAMP and found label located in both the NH_2- and COOH-terminal halves of the protein after cleavage with chymotrypsin.

In summary, the location of the site of labeling by a photoaffinity reagent within a polypeptide is similar in principle to locating residues modified by other means. There are, however, a few important differences that must be kept in mind.

(i) Serious problems may arise if the stoichiometry of labeling is low. Methods for obtaining high extents of labeling are given below.

(ii) Labeled polypeptides may behave differently in several ways when compared with unlabeled polypeptides. They may have altered cleavage patterns with both chemical reagents and enzymes. The peptides derived from them will usually separate from the unlabeled peptides in chromatographic systems with the exception of gel filtration. Edman degradation may be blocked at the modified sites.

(iii) The linkages between the reagent and the receptor may be labile under extremes of pH, in the presence of strong nucleophiles, etc. For example if an aryl carbene reacted with the carboxyl group of Glu or Asp a benzyl ester, labile to hydrolysis, nucleophiles and catalytic reduction, would be formed. (See Abercrombie et al., 1982; Mas et al., 1980; Fisher and Press, 1974; for examples of labile linkages.)

(iv) Even if the region of the receptor which is labeled is restricted, numerous radioactive peptides can be formed because photogenerated reagents may react with many different bonds.

A powerful method for identifying labeled peptides and residues that may be developed in the future is combined gas chromatography–mass spectrometry. After partial acid or enzymatic digestion and suitable derivatization the mixture of labeled and unlabeled fragments would be separated by gas chromatography and the mass spectrum of each peak recorded (see e.g. Kelley et al., 1975). The resolving power and sensitivity of the technique is so great that most of the problems listed above may be reduced in significance. Lindemann and Lovins (1976 and references therein) pioneered the application of mass spectroscopy for the identification of photochemically derivatized amino acids but there are several deficiencies in their data and new work in this area is needed.

4.6. The extent of labeling

When a receptor in a mixture of molecules is identified by photoaffinity labeling, the stoichiometry of labeling is unimportant as long as a specifically labeled polypeptide is identified. As we have seen above, however, if the intention is to define the labeling site more precisely a high extent of labeling is most helpful. The extents of labeling that have been noted in the literature vary enormously. In several cases an almost quantitative modification of the binding site has been achieved (e.g. Walter et al., 1977; Rangel-Aldao et al., 1979). In other cases yields have been high (e.g. Forbush et al., 1978: 30 to 40% labeling of Na,K-ATPase with a cardiotonic steroid derivative), moderate, or poor (e.g. Niedel et al., 1980: 0.1% labeling of chemotactic growth factor receptor; Aiba and Krakow, 1980: 2.5% labeling of a cAMP binding protein).

There is no clear-cut relationship between the nature of the ligand and the extent of labeling. One cause of low extents of labeling is destruction of the binding site of the receptor. This often occurs when a reagent with a low quantum yield for attachment or a low extinction coefficient is used, so that photochemical damage to the receptor occurs before much labeling has taken place. When Forbush and Hoffman (1979) labeled Na,K-ATPase with unmodified cardiotonic steroids < 1 % labeling could be achieved before photocrosslinking of the enzyme became severe. When aminoacyl tRNA synthetases were photocrosslinked to unmodified ATP the extent of labeling was limited to 15 % by photochemical damage to the nucleotide binding site (Yue and Schimmel, 1977). Preirradiated enzyme would not incorporate ATP. Destruction of a binding site might also be caused by ligand dependent reactions including sensitized photooxidation and other photochemical reactions that do not result in covalent attachment (Section 4.4).

A second cause of non-quantitative labeling is poor efficiency of covalent attachment of the activated species at the binding site. This may occur if the photogenerated intermediate rearranges to an inert species, if the intermediate can react with water or a buffer component near the binding site because it is poorly oriented for reaction with the protein, or if the intermediate is long-lived enough to dissociate from the active site and react with the surrounding medium or other components of the system under study (see the discussions of non-specific labeling and pseudo-photoaffinity labeling below). Poor orientation may be a particular problem with photolabile derivatives of macromolecules (Fig. 3.26).

A third cause can be the tight binding of photolysis products derived from the reagent to the ligand binding site displacing unphotolysed molecules (see Maassen and Moller, 1978, and Standring and Knowles, 1980 for examples of photolysis products that bind tightly).

The fraction of labeled receptor in a sample may be increased in two ways: either the labeling experiment can be repeated several times on the same sample thereby increasing the extent of labeling, or the covalent ligand–receptor complex can be separated from unlabeled receptor. Cooperman and Brunswick (1973) have provided an extensive discussion of the former approach. They found that the extent of labeling of phosphofructokinase with a diazomalonyl derivative of cAMP saturated at 35 % unless the labeling experiment was repeated after the low molecular weight photolysis products were removed, by dialysis or by washing an ion-exchange column to which the enzyme was bound. If labeling was repeated up to seven times more than 70% of the receptor sites could be modified. The dialysis method (or for membranes, centrifugation and resuspension) followed by relabeling has been used successfully by many workers. It may fail to produce the desired increase in labeling if the photodecomposition products of the ligand bind so tightly to the receptor that they cannot be removed under non-denaturing conditions (e.g. Fisher and Press, 1974; Standring and Knowlss, 1980). Where hydrophobic ligands are involved it is often helpful to include bovine serum albumin in the washes.

The separation of labeled and unlabeled receptors can best be achieved by affinity chromatography after the removal of unbound ligand. This method has the advantage that it removes both unlabeled and non-specifically labeled molecules. Fisher and Press (1974) labeled antibodies against the 2-nitro-4-azidophenyl group and then isolated the protein that did not stick to a 2,4-dinitrophenyl–Sepharose column. Lau et al. (1977) labeled acetyl CoA–glycine *N*-acetyltransferase with *p*-azidobenzoyl-CoA and separated the labeled and unlabeled enzyme on Blue dextran–Sepharose. Tightly but non-covalently bound photolysis products can interfere with such procedures. Isoelectric focusing (Swanson and Dus, 1979; Brems and Rilling, 1979) and ion-exchange chromatography (Havron and Sperling, 1977) may also prove useful for isolating ligand–receptor covalent adducts, especially if the ligand has a net charge.

Where a reagent binds very tightly to a receptor before irradiation it is possible to correlate the extent of labeling with the initial concentration of reagent that is used and so obtain ligand–receptor association constants (Walter et al., 1977; Skare et al., 1977; Rangel-Aldao et al., 1979). For example, the association constants for receptors represented by labeled bands on SDS-polyacrylamide gels may be estimated in this way. The inhibition of labeling by a non-photoactivatable ligand may also provide useful information in this regard (Walter et al., 1977; but see Pomerantz et al., 1975). However in more complex, but nevertheless typical, situations there may not be a simple relationship between reagent concentration and the extent of labeling (Yue and Schimmel, 1977; Munson and Kyte, 1981; and see below).

Consider, for instance, the case where a photoaffinity reagent, L, binds to a receptor, R, and the dissociation constant is K. On photolysis either a covalent complex R–L or a photolysis product L' is formed which has a dissociation constant, K'.

$$
\begin{array}{ccccc}
R+L & \overset{K}{\rightleftharpoons} & R\cdot L & \xrightarrow{k_1} & R-L \\
\downarrow k' & & \downarrow k'-k_1 & & \\
R+L' & \overset{K'}{\rightleftharpoons} & R\cdot L' & &
\end{array}
$$

If $[R]_{\text{Total}}$ is small (i.e. $[L]_0 \simeq [L] + [L'])$ then as $t \rightarrow \infty$, it can be shown that:

$$
\frac{[R-L]}{[R]_{\text{Total}}} = 1 - \left(\frac{K([L]_0 + K')}{K'([L]_0 + K)}\right)^{\frac{k_1}{k'} \cdot \left(\frac{K'}{K'-K}\right)}
$$

For small K' (photolysis products bind strongly):

$$
\frac{[R-L]}{[R]_{\text{Total}}} = 1 - \left(\frac{K[L]_0}{K'([L]_0+K)}\right)^{\frac{-k_1}{k'} \cdot \frac{K'}{K}}
$$

For large K' (photolysis products bind weakly):

$$
\frac{[R-L]}{[R]_{\text{Total}}} = 1 - \left(\frac{K}{K+[L]_0}\right)^{\frac{k_1}{k'}}
$$

For $K' = K$:

$$
\frac{[R-L]}{[R]_{\text{Total}}} = 1 - \mathrm{e}^{\frac{-k_1}{k'}\left(\frac{[L]_0}{[L]_0+K}\right)}
$$

TABLE 4.2

Calculation of $[R\text{–}L]/[R]_{Total}$ after prolonged irradiation of a photoaffinity reagent, L, a receptor, R. (Symbols are defined in the text.)

K'	Value[a] of $[R\text{–}L]/[R]_{Total}$		Ratio[b] of values for
	$[L]_0 = 0.1$	$[L]_0 = 100$	$[L]_0 = 0.1$ and 100
10	0.188	0.874	0.215
0.1	0.139	0.259	0.537
0.001	0.012	0.014	0.857

[a] For $K = 0.1$; $k_1/k' = 0.3$

[b] As the value of $[R\text{–}L]/[R]_{Total}$ for $[L]_0 = 100$ can be taken to be the maximum fraction of the receptor that can be labeled, this ratio represents the extent of labeling expressed as a fraction of the maximum for $[L]_0 = K = 0.1$.

To illustrate the point let us consider the case where $K = 0.1$, the labeling efficiency $k_1/k' = 0.3$, $[L]_0$ (the initial reagent concentration) is either 0.1 $(=K)$ or 100, and calculate the ratio of the extent of labeling at these two concentrations at long irradiation times (Table 4.2).

As expected, for both values of $[L]_0$, the maximal extent of labeling drops if the photolysis product L' binds strongly to R. Perhaps surprisingly, the extent of labeling when $[L]_0 = K = 0.1$, and half the receptor sites are occupied initially is higher than half-maximal (using $[L]_0 = 100$ to obtain the maximal extent), when L' binds strongly, demonstrating that K cannot be derived simply from the concentration of reagent required to produce half-maximal labeling.

Munson and Kyte (1981) used a similar analysis in a study of the labeling of Na,K-ATPase with an ATP analog. Of course the kinetic scheme can be complicated in a number of ways, for instance the rate of photolysis may be fast compared with the rate of attainment of the equilibria depicted.

4.7. *Control experiments*

Certain control experiments are essential to all studies using photoaffinity reagents. In particular, the following should be tested: the stability of the receptor to irradiation, the stability of the ligand including its reactivity towards the receptor in the dark, and the occurrence of non-specific labeling, that is labeling outside the ligand binding site. There are useful ways to detect and prevent the latter.

4.7.1. Stability of the receptor

It is important to test the stability of the biological sample to the radiation doses that are required for labeling. The unpleasant consequences of destroying a binding site faster than it is labeled, or of photochemically crosslinking receptor molecules to others can easily be imagined. As indicated earlier, in some cases it may be advantageous to irradiate the reagent at 254 nm, but then it is often necessary to control carefully the photolysis time to avoid complications. At 300 nm and above most receptor preparations are rather stable. Of course, if a component such as a cytochrome absorbs at higher wavelengths care should be taken to avoid strong absorption bands (e.g. Swanson and Dus, 1979).

The stability of the sample may be tested in several ways. Functional assays of the receptor under investigation should be done after irradiation in the absence of the reagent. If a purified low molecular weight protein is the target of study, amino acid analysis may be used to determine the extent of destruction of susceptible amino acids such as Trp and Cys (see Cooperman and Brunswick (1973) for a discussion and for an example see Galardy et al., 1974). In more complex samples, such as an intact membrane preparation, photochemical damage may become apparent when gel electrophoresis is carried out. Both crosslinking of proteins (and other components such as nucleic acids), and occasionally cleavage of polypeptide chains may occur. The susceptibility of polypeptides to photochemical damage is extraordinarily variable. On one extreme, chloroplast membranes are resistant to damage, which might be expected, while in contrast, acetylcholinesterase from *E. electricus* appears to be extremely sensitive to destruction by irradiation in the 280 nm region. Cleavage of the polypeptide chain and inactivation of the enzyme occur simultaneously (Bishop et al., 1980). Irradiation at longer wavelengths and the use of an inert atmosphere are often useful in preventing the photochemical destruction of susceptible molecules.

4.7.2. Stability of the reagent

Two distinct problems can arise with the stability of the photoaffinity reagent itself and both of these have been alluded to earlier. First, the reagent may be unstable under physiological conditions. Several such cases have been mentioned in Chapter 3 including the instability of aryl azides and diazo compounds towards thiols, and the instability of diazoacetyl compounds at low pH. Reagents may also be altered enzymatically for example endogenous phosphodiesterases can attack cAMP derivatives (e.g. Wallace and Frazier, 1979). To ensure that such transformations are not occurring a control experiment in which the integrity of the reagent is directly assayed should be done. It is most satisfactory to perform a chromatographic analysis (TLC or HPLC) on the reagent after incubation with the biological preparation. This is usually simple if a radiolabeled reagent is being used. Unlabeled cold carrier may be added and the reagent extracted into organic solvent, or the protein in the sample is precipitated (e.g. by the addition of trichlo acetic acid or acetone) when the reagent remains in the supernatant. A portion is then tested to see if the molecule has remained intact. It may be more difficult to recover and test the integrity of macromolecular reagents. The stability of the photoactivatable group can be assayed though, by using a low molecular derivative. For example, if a reagent was made by coupling 2-nitro-4-azidophenyl fluoride to a polypeptide and it was used under conditions where the azido group might become reduced, the stability of *N*-(2-nitro-4-azidophenyl)glycine could be tested under the conditions prevailing in the reaction mixture.

The second problem related to reagent stability is covalent attachment in the dark, either specific or non-specific, to components of the preparation under study. All the advantages of photoaffinity labeling are lost if this occurs. On the whole, reaction before photolysis has not been a serious problem with photoaffinity reagents except diazoacetyl derivatives which were indeed among the earliest chemical affinity reagents (see Baker, 1964). The modification of histidine in diazoacetylchymotrypsin probably occurs in an acid-catalysed rather than a light-induced process (Shafer et al., 1966). Another seemingly unlikely variation has been noted. After performing an extensive photoaffinity labeling study with epidermal growth factor derivatives (Das and Fox, 1978), Fox's group found conditions under which iodinated but otherwise unmodified EGF becomes covalently attached to its receptor in a spontaneous reaction in the dark (Linsley et al., 1979). It was claimed that in the earlier experiments, which were performed at a lower temperature, no attachment occurred in the dark and true photoaffinity labeling occurred on photolysis. In a similar case it was found that α-thrombin derivatives and a receptor on mouse embryo cells couple in the dark (Carney et al., 1979). At first, these events were thought to be of biological significance. Very recently, however, Comens et al. (1982) have shown that the dark reaction may be an artefact of chloramine-T iodination, at least in the case of epidermal growth factor.

4.7.3. Non-specific labeling

One of the most important considerations in photoaffinity labeling is the detection and prevention of non-specific labeling. Non-specific labeling is that which occurs outside the ligand binding site, either within the receptor or with completely unrelated molecules. Non-specific labeling can arise if an unnecessarily high concentration of reagent is used. In this case both the true receptor site and weak binding sites unrelated to the function of the ligand will be labeled. The lipid bilayer of membranes is considered as a weak binding site for hydrophobic molecules, and this must be taken into account when labeling cell-surface receptors. Further, random labeling will occur through bimolecular reactions between components of the sample and activated reagent molecules. The latter form of non-specific labeling will be a particular problem when attempts are made to label relatively weak binding sites, e.g. $K_d > 10^{-5}$ M by using high concentrations of ligand to obtain high occupancy. Non-specific labeling may also occur when K_d is low, but association and dissociation is fast, if the binding site is relatively inert. Although many photogenerated intermediates are highly reactive (Chapter 2) most exhibit substantial selectivity and it is quite conceivable that an activated ligand might dissociate from a binding site that is hydrocarbon in character and couple with a highly reactive residue outside the site, for instance a nucleophilic sulfhydryl group. If the photogenerated intermediate binds less well to the receptor than the unphotolysed ligand the situation will be exacerbated. When an activated ligand is so inert that it can essentially equilibrate between its binding site and the aqueous phase many of the advantages of photoaffinity labeling are lost and this situation which resembles a conventional affinity labeling experiment has been termed pseudophotoaffinity labeling (Ruoho et al., 1973; see also Smith and Knowles, 1974; Bayley and Knowles, 1977; and Westheimer, 1980 for further discussion).

In conventional affinity labeling, the usual way to distinguish between specific and non-specific labeling is to use a protecting reagent, usually the natural ligand. The same is done in photoaffinity labeling. The reagent will be displaced from the receptor and specific labeling will be prevented. While specific labeling should be saturatable, non-specific labeling normally rises almost linearly with reagent concentration (see Section 4.6 and Munson and Kyte (1981) for an analysis). Where both are occurring it should be possible to subtract the values for label incorporation, with and without protection, at various initial reagent concentrations to obtain a

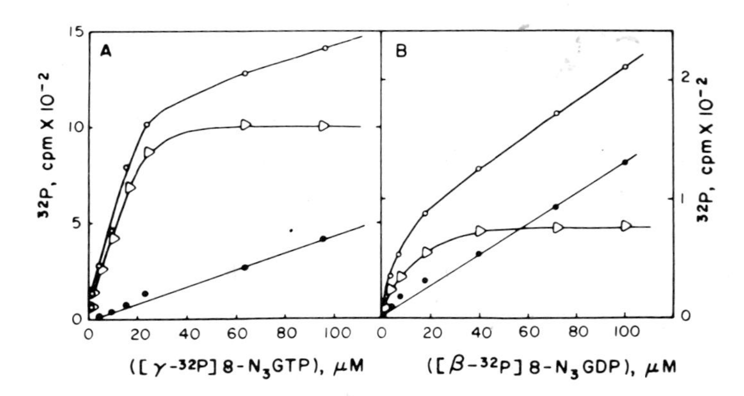

Fig. 4.5. Photoaffinity labeling of tubulin with [γ-^{32}P]8-N_3GTP and [β-^{32}P]8-N_3GDP. Tubulin samples (0.5 mg/ml) in reassembly buffer (pH 6.7) were extracted with charcoal and incubated at 4 °C for 5 min with varying concentrations of [γ-^{32}P]8-N_3 GTP (300 cpm/pmol) or [β-^{32}P]8-N_3GDP (95 cpm/pmol). The solutions were separated by sodium dodecyl sulfate-polyacrylamide gel slectrophoresis and the incorporation into each band (α or β) was quantified. (O) Total incorporation into β-monomer; (●) total incorporation into α-monomer; (Δ) incorporation into β-monomer minus incorporation into α-monomer for [γ^{32}P]8-N_3GTP (A) and [β-^{32}P]8-N_3GDP (B). (From Geahlen and Haley, 1979.)

saturation curve. Non-specific labeling of the α-subunit of tubulin by 8-N_3-GTP or 8-N_3-GDP was clearly observed by Geahlen and Haley (1979; Fig. 4.5). The β-subunit was labeled both specifically and non-specifically in these experiments (Fig. 4.5). The investigators substracted the values for the labeling of α from those for β at each reagent concentration to obtain values for specific labeling of β. It was fortuitous that the non-specific labeling of α and β had similar concentration dependences. A similar result should have been obtained by the protection experiment suggested above which is generally valid, or by subtracting from the curve for the labeling of β a straight line through the origin fitted to the slope of the curve at high reagent concentrations (see also Mohler et al., 1980). Figure 4.5 also illustrates the trade-off between a high extent of specific labeling and a relative increase in non-specific labeling.

Protection should be immediately obvious when the results of a labeling experiment are examined by gel electrophoresis. A particularly clear-cut example is reproduced in Fig. 4.6 (Pomerantz et al., 1975; see also Fig. 4.4). Another instructive example can be seen in the work of Nordeen et al. (1981) in which specific labeling of glucocorticoid receptors was distinguished from a strongly but non-specifically labeled component of cell cytosol.

If the photoaffinity reagent has been designed well and there are a measurable number of tight binding sites for it, the protection experiment will usually succeed. Occasionally it will not; for example, four polypeptides of sarcoma cells were labeled with *low* concentrations of 8-N_3-cAMP but only three of the sites could be protected by cAMP. Presumably, the fourth site is not part of a cAMP binding protein but a different nucleotide binding site that just happens to have affinity for the reagent but not for cAMP.

Another test for specific labeling is to determine whether the ligand binding site is blocked. But, as I pointed out in the discussion of photoinactivation experiments, there are several other possible causes of apparent binding site occupation besides the covalent attachment of a ligand. It has also been noted that specific labeling as defined by a protection experiment may not always yield a blocked receptor when the photoaffinity label is a macromolecule. When sodium channels in tissue culture cells were labeled

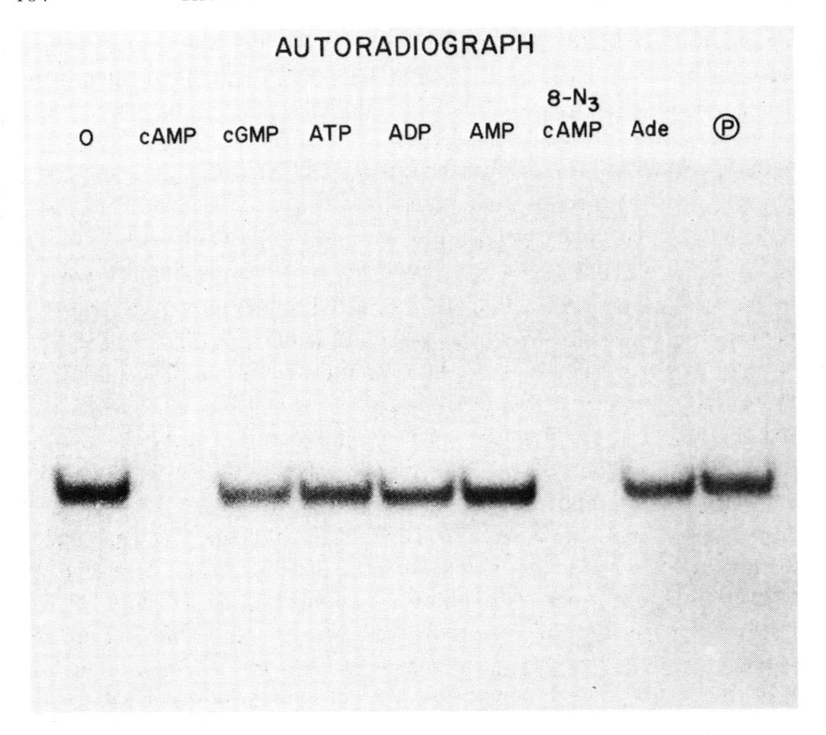

Fig. 4.6. Autoradiograph of a sodium dodecyl sulfate-polyacrylamide gel showing the selective incorporation of 8-N_3-cAMP into protein kinase regulatory subunit. Protein kinase, prior to irradiation, was incubated with 0.2 μM [^{32}P]8-N_3-cAMP in 50 mM sodium morpholinoethanesulfonate (pH 6.2), with 10 μM amounts of the indicated unlabeled nucleotides or adenosine (Ade). In the last lane of the gel was the regulatory subunits of the protein kinase which was endogenously phosphory lated with [γ-^{32}P]ATP. Each lane of the gel contained 40 μg of total protein. The amounts of [^{32}P]8-N_3cAMP incorporated into the regulatory subunit were (in fmoles per lane): with no addition 110; + cAMP, <1; + ATP, 93; ADP, 93; + 5′-AMP, 106; + unlabeled 8-N_3-cAMP, <1; + adenosine, 77. (From Pomerantz et al., 1975.)

with an arylazido derivative of ^{125}I-labeled scorpion toxin, 50 to 70 % of the specifically bound label became covalently attached yet only 17 % of the toxin binding sites were blocked (Beneski and Catterall, 1980). The authors suggested that arylazido groups attached at sites on toxin molecules remote from the binding domain might become covalently bound to channel components in such a manner that the bound toxin can be apparently displaced by unmodified toxin (Fig. 4.7: and see the discussion of endo- and exo-reagents in Section 3.6.1). ·

The detection of non-specific labeling *within* a given polypeptide may also be detected by protection experiments. It would be most satisfying to demonstrate that specific peptide fragments are protected from labeling by the natural ligand but such experiments have not been done. In a revealing example, however, Lifter et al. (1974) labeled a mouse myeloma IgA with 2,4-dinitrophenylazide. Reaction occurred at two tyrosine residues believed to be too far apart (~ 23 Å) to be at the same site. Experiments in which the extent of labeling of a polypeptide is measured in the presence and absence of a protecting agent have provided useful information on non-specific labeling within the receptor polypeptide. In a particularly useful set of publications the labeling of dehydrogenases with three NAD analogs (the diazoacetate ester of the 3-(hydroxymethyl)pyridine analog (Browne et al. 1971), the 3-azidopyridine analog (Hixson and Hixson, 1973), and more recently the 3-diazirino analog (Standring and Knowles, 1980) has been described. In the last case, by performing protection experiments with NAD and by using scavengers (see below), it was demonstrated that after labeling at ~ 25% site occupancy and gel filtration, the label that remained associated with the protein was distributed as follows: specific covalent labeling, 8% of the sites; non-specific covalent labeling, 3% of the sites, and non-covalent association (tight binding photolysis products at active site released on denaturation), 5% of the sites.

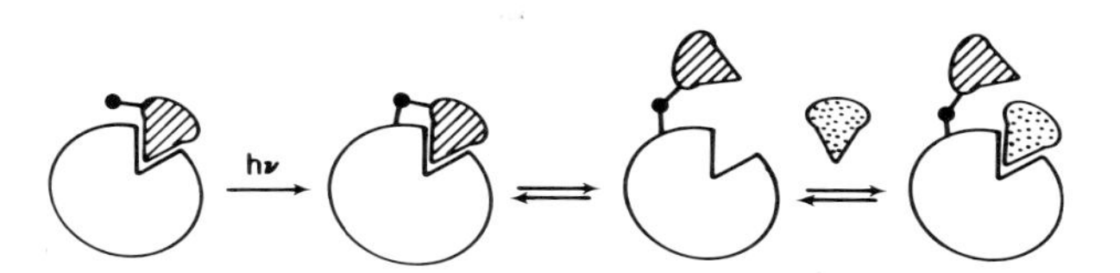

Fig. 4.7. Ligands may still bind to a receptor after covalent modification with a photoaffinity reagent. Here a receptor (unshaded) is labeled with a macromolecular reagent (hatched) made by attaching a bifunctional reagent to a natural ligand (stipled). For an explanation see the text.

As the dissociation constant, and perhaps the dissociation rate, of the diazirino analog of NAD was high (4 mM) the high extent of specific labeling was attributed to the high reactivity of the carbene generated at the active site which reacted before dissociation followed by reaction elsewhere could occur. A greater relative extent of non-specific labeling was observed with the other analogs presumably because less reactive photolysis products are formed from the diazoacetyl and arylazido groups (see Chapter 2). The linear diazo isomer formed from the diazirine appeared to be inert under the prevailing conditions.

The properties of an ideal protecting agent deserve some discussion. Of course it should bind tightly to the receptor. The concentration that must be used can be predicted from the K_d values for the photoaffinity reagent and the protecting molecule, if exchange in and out of the binding site is slow relative to the time required for photolysis. Where exchange can occur, labeling is inefficient, and photolysis products bind to the receptor, the situation is complex and these K_d values are only an approximate guide to the concentration of protecting agent that must be used (e.g. Pomerantz et al., 1975). For example, where the photolysis products bind weakly and labeling is inefficient, half-maximal labeling occurs at ligand concentrations well above the K_d measured in the dark (Table 4.2). Thus a higher concentration of protectant than that required to prevent half the site occupuncy in the dark would be required to prevent half-maximal labeling.

Ideally, the protecting agent should not absorb light and so screen the reagent, reducing the extent of photolysis in a given time. If tight binding sites are under investigation protection may be achieved with a concentration of protecting agent that does not affect light absorption by the reagent, but where micromolar concentrations are required for protection, further control experiments are necessary. The use of a split cell is a possibility but this will overcompensate for the absorption of the protecting agent (Ruoho et al., 1973). Perhaps the most convincing approach is to show that a substance that is not a ligand but does absorb radiation to the same extent as the protecting agent cannot protect the receptor from labeling (see Fig. 4.6). An alternative is to demonstrate, using an appropriate analytical method, that the reagent was photolysed in the presence of the protecting agent. Another criterion for an ideal protecting agent is that it should not scavenge reactive intermediates (see below).

Chen and Guillory (1979) have presented a variation on the protection experiment which should be generally useful where a reagent has been made by coupling a photoreactive group to a natural ligand. They used an NAD derivative in which 3-[*N*-(4-azido-2-nitrophenyl)amino]propionic acid had been coupled to a hydroxyl of the adenosine. This molecule irreversibly inhibited NADH-CoQ reductase on irradiation but NAD and the propionic acid derivative together had no such effect, i.e. photolysis with the aryl azide was not damaging the enzyme in some non-specific manner. In another instructive case it was shown that inactive insulin analogs were poor at protecting the insulin receptor from labeling, i.e. the possibility that insulin itself, which was effective, scavenged the photoactivated derivative was ruled out (Yeung et al., 1980). Wallace and Frazier (1979) found that a protein of *D. discoideum* membranes labeled with 8-N_3-cAMP was not protected by cAMP but it was protected by AMP. In fact, 8-N_3-AMP had been formed by an endogenous diesterase and it labeled actin.

4.7.4. The use of scavengers

Scavengers may be used to eliminate non-specific labeling and pseudophotoaffinity labeling. The principle is illustrated in Fig. 4.8. A molecule which is capable of trapping reactive intermediates, including the primary photolysis products and longer lived rearrangement products, is added to the buffer in which the sample is contained. Photogenerated intermediates formed outside the ligand binding site, or those that migrate out of the binding site after being formed there, react with the scavenger and non-specific labeling is prevented. Pseudophotoaffinity labeling, in which relatively long-lived photogenerated species exchange in and out of the binding site before reacting there is also eliminated. In some cases this labeling mechanism gives rise to a substantial fraction of the specific labeling (e.g. Payne et al., 1980).

Many reagents have been proposed as scavengers. Ideally, the scavenger should not absorb light and of course it should not bind to the receptor.

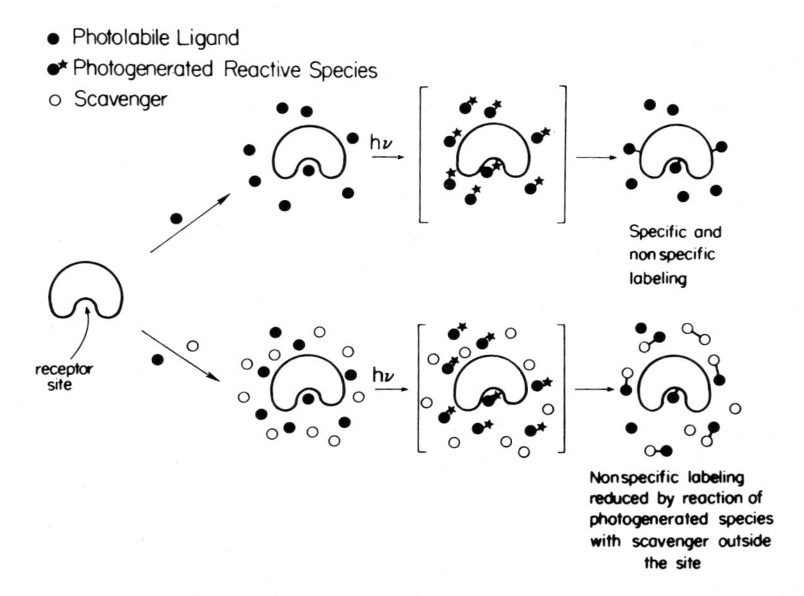

Fig. 4.8. The use of scavengers to increase the proportion of specific binding site labeling. (From Bayley and Knowles, 1977.)

Amines (*p*-aminobenzoic acid, *p*-aminophenylalanine, Tris), hydroquinone, thiols and soluble proteins have been used (see Bayley and Knowles, 1977 and references therein). The rationale behind using a protein is that it will contain all the functional groups that might be available in the biological preparation and will compete effectively with them for reactive species outside the binding site. If a protein scavenger is used it should be ensured that it does not bind the ligand: for example bovine serum albumin would be a poor choice if a lipophilic reagent were being used.

Although no definitive study of scavengers in photoaffinity labeling experiments has yet been made, the present indications are that thiols are the most suitable reagents. It is expected that as strong nucleophiles they should react with most photogenerated species which are electrophiles as well as with free radicals which can abstract hydrogen atoms from the weakly bonded —SH group. Care must be taken to ensure that the thiol does not react with the reagent in the dark (Chapter 3).

Bayley and Knowles (1978a,b 1980) used 15 to 50 mM reduced glutathione to scavenge reactive intermediates generated from diazirines and aryl azides. Standring and Knowles (1980) made a careful study of the non-specific labeling of lactate dehydrogenase with a diazirino analog of NAD. Here 5 to 25 mM glutathione was effective in scavenging carbenes outside the NAD binding site. β-Mercaptoethanol, thiolactate, and thioglycolate were also effective. It is noteworthy that Tris which has been proposed as a scavenger was ineffective in this case. Payne et al. (1980) also used β-mercaptoethanol (20 mM) to scavenge a long-lived intermediate formed on photolysis of a diazoketone. The half-life of the scavengeable species was so long ($T_{1/2}$ 17 h) in the absence of β-mercaptoethanol that it was unlikely to have been a ketene and it may have been the product of a reaction between the Wolff rearrangment product and the nucleophilic azide anion present in the buffer as an antibacterial agent (J. Katzenellenbogen, personal communication: Fig. 4.9). Obviously, when reactive intermediates are so long-lived, immediate washing after photolysis can be helpful.

In a dramatic demonstration of the use of a scavenger Nicolson et al. (1982) demonstrated that a long-lived intermediate formed when *p*-azido[^{3}H]puromycin was used to label *E. coli* ribosomes was diverted from its reaction with polypeptide S18 by 2 mM β-mercaptoethanol. S18 was the

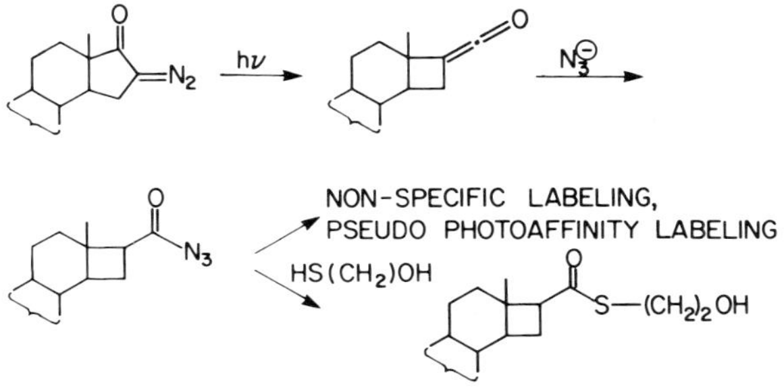

Fig. 4.9.

major labeled polypeptide in the absence of a scavenger and it is likely that the highly reactive Cys-10 residue was attacked by a free reactive intermediate.

As no systematic study has been made of the concentration of a thiol required to scavenge various reactive species a study of the concentration dependence should be made in each case to ensure that maximal prevention of non-specific labeling is obtained. It should not be assumed that the efficacy of a thiol scavenger will be independent of pH as thiols have p*K* values of ~ 9.

In summary, specific labeling of binding sites by photoaffinity reagents can be brought about, while reducing the extent of non-specific labeling, by paying attention to a number of factors. These include the proper design of the reagent (it must bind tightly), the manipulation of receptor and ligand concentrations to maximize the ratio of bound : free reagent, and irradiation for just enough time to photolyse the reagent (too long a time may lead to cleavage of the ligand–receptor bonds or to slow non-specific labeling by radioactive photolysis products). If the rate of exchange of ligand at the binding site with free ligand is slow it may be possible to remove unbound ligand before irradiation. Where this is not possible the most reactive photogenerated intermediate (most short-lived) should be used to prevent pseudophotoaffinity labeling and related problems. Non-specific labeling may be detected with agents that protect the binding site and be prevented with scavengers.

4.8. Some miscellaneous methods

Some miscellaneous methods which may grow in importance deserve a brief review. The potential use of photogenerated reagents in experiments where time-dependent phenomena are examined was mentioned in Chapter 1. Many methods for investigating rapid events in biological systems, such as the use of fluorescent probes, yield little structural information. Examples of experiments with photoactivatable reagents that do yield such information have appeared recently. They include the use of a hydrophobic reagent to monitor the interaction of cholera toxin with membranes

(Wisnieski and Bramhall, 1981: Section 6.3) and a study of how RNA polymerase finds a promotor site (Park et al., 1982a,b; Section 1.2.5). The experiments of DeRiemer and Meares (1981) on the interaction of a growing RNA chain, with a 5′-photoactivatable group with the subunits of RNA polymerase are also instructive. These experiments were performed in a stepwise rather than a time-dependent manner. In time-dependent experiments it is crucial that the photogenerated intermediates short-lived compared with the half-life of the species under study, and this matter requires careful consideration (Sections 2.3, 3.2.4, 3.3.4, 3.4.4, 4.7.3, 4.7.4).

Another interesting possibility is to carry out the labeling experiments at low temperatures to characterize transient biological intermediates, and to prevent non-specific labeling by reagent molecules that dissociate from the binding site after activation. The success of such experiments will depend on the activation energy of the rate-determining step of the labeling reaction being low compared with those for the other processes under consideration. Marinetti et al. (1979) irradiated bacterial photosynthetic reaction centers, reconstituted with 2-azido[^{3}H]anthraquinone, at 80 K to yield a nitrene which was stable until the sample was slowly warmed when covalent attachment to a protein subunit occurred. When the irradiation was performed at 25°C, extensive denaturation of the protein occurred, with little covalent attachment of the reagent.

Photoaffinity labeled sites have been visualized in thin tissue sections by autoradiography (Mohler et al., 1980, 1981). Immunochemical methods have also proved useful. Olson et al. (1982) labeled ribosomes with puromycin and examined the sites of attachment in the electron microscope after reacting the covalently bound ligand with anti-*N*,*N*-dimethyladenine antibodies. Gronemeyer and Pongs (1980) visualized ecdysterone photochemically attached to polytene chromosomes, by indirect immunofluorescence. Schaltmann and Pongs (1982) further demonstrated that polypeptides labeled by ecdysterone could be detected immunochemically in gels by a Western blot procedure. Several fluorescent photochemical reagents have been made (e.g. Dreyfuss et al., 1978; and Section 6.1).

CHAPTER 5

Photochemical crosslinking reagents

This chapter is a summary of a field that is still under development: crosslinking with photochemical reagents. In a photochemical crosslinking experiment a photoaffinity reagent need not be specially prepared. Instead a generally useful reagent with two photoactivatable groups or a reagent with *one* chemically reactive group and one photoactivatable group is added to a preparation of the structure that is under investigation and, after appropriate incubation, the preparation is irradiated. The investigator is usually interested in protein–protein crosslinks which can provide information on the subunit structure of a protein or a nearest neighbor analysis of proteins in more complex structures such as ribosomes or membranes. Conventional chemical crosslinking has proved useful in these areas as well as in investigations of muscle, virus and chromatin structure. Peters and Richards (1977) have given a thoughtful account of conventional reagents and the early work on photochemical crosslinkers. Photochemical reagents in particular have been reviewed by Das and Fox (1979) and by Ji (1979).

5.1. The advantages of photochemical reagents

The advantages of photochemical reagents in general have been described earlier and they apply to crosslinking reagents too. However, the possibility of triggering the reagent at will has for the moment been partly negated by the necessity of using reagents with one photochemically activated arm and one conventional chemical arm (see below). The high reactivity of a photochemical reagent is important as it allows the formation of crosslinks that might not occur with group specific reagents. For example, if dimethylsuberimidate becomes attached by one of its arms to one protein and there is no amino group available for reaction on a neighboring protein, the second arm will react with water and be inactivated. A photogenerated species such as a nitrene or a carbene would have a better chance of forming a covalent crosslink. This exemplifies the important point that the failure of a reagent to form a crosslink between two polypeptides does not necessarily mean that they are not neighbors. An eventual goal in the field of photochemical crosslinking is to develop reagents so indiscriminate in their reactions that the failure to crosslink two components of a biological system actually provides useful information.

The reactivity of a reagent is related to its half-life in a biological system which is an important factor under certain circumstances. For instance, in biological membranes it is not a simple matter to distinguish between stable complexes and transient complexes formed by collisions between the highly concentrated and rapidly diffusing proteins in the bilayer. A more or less permanent neighbour may be inert towards the second arm of a crosslinking reagent while a momentary encounter with a protein containing a highly reactive group might result in covalent attachment. Again, it is hoped that highly reactive photochemical reagents may be designed to eliminate this problem.

5.2. Reagents with two photoactivatable groups

Hydrophobic crosslinking reagents each with two photochemically activatable groups have been synthesized by Mikkelson and Wallach (1976). Their reagents (Fig. 5.1) were designed to bind to the lipid bilayer of

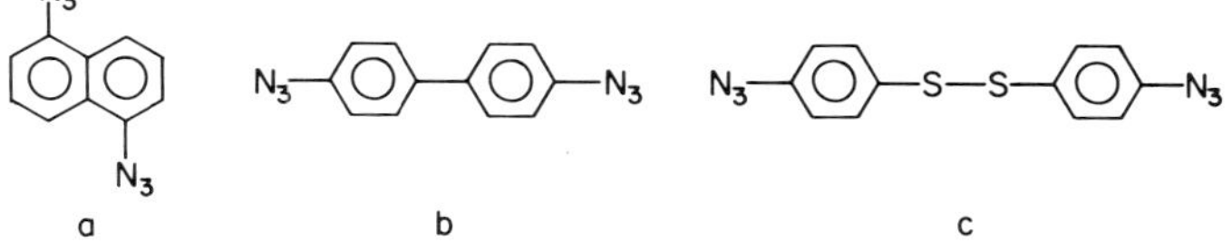

N_3

N_3

a

N_3—⟨benzene⟩—⟨benzene⟩—N_3

b

N_3—⟨benzene⟩—S—S—⟨benzene⟩—N_3

c

Fig. 5.1.

membranes and crosslink within it. Such reagents have the obvious disadvantage that if the efficiency of attachment is low for each activated azido group, say 10% (see Chapter 6), then the extent of crosslinking will be very low: 1%. The results of Mikkelson and Wallach cannot be clearly interpreted. Using the cleavable reagent (Fig. 5.1c) appreciable crosslinking was achieved but much of it could not be reversed by β-mercaptoethanol. Further, these investigators used 0.1 mM DTT in their photolysis buffer which, if it penetrated the bilayer, could have reduced both the azido groups and the disulfide linkage of the reagent before irradiation. Spectrin, a peripheral protein, was unexpectedly crosslinked. A useful control would have been to see whether a hydrophobic monoazide induced crosslinking under the photolysis conditions. A definitive investigation of reagents with two photoactivatable groups for protein crosslinking is required.

Reagents with two photoactivatable groups may be more useful in situations where tight binding to the target and very sensitive detection techniques are available. Mitchell and Dervan (1982) have described a reagent in which two monoazidomethidium molecules are separated by a polyoxyethylene chain 25 Å in length. On irradiation with phage λDNA, 6% of the reagent was attached, resulting in DNA crosslinking with an efficiency of 0.3 to 0.4% based on the amount of reagent used. After treating DNA with 1 crosslinker per 100 base pairs, crosslinking was easily observable in the electron microscope. The utility of psoralens for photochemical nucleic acid crosslinking was mentioned earlier (Section 2.3.7) and bis-psoralens for interhelical crosslinking have been synthesized (Schwartz et al., 1983).

5.3. *The design of photochemical crosslinking reagents*

The reagents in current use for protein crosslinking have one chemically reactive and one photochemically reactive arm, often connected by a cleavable bridge for subsequent two-dimensional electrophoretic analyses. Chemical attachment is carried out first, and the crosslink is completed by photolysis. Many of these so-called heterobifunctional molecules are listed in Table 5.1 and some of them, conforming to the criteria outlined below, are suitable for photochemical crosslinking studies, although many were designed with the primary purpose of attaching them to macromolecules to form photoaffinity reagents (Section 3.6).

Important factors in the design or choice of a heterobifunctional photochemical reagent are the nature of the chemical and photochemical arms, the solubility and permeability properties of the molecule, the nature of the cleavable bridge between the two arms, and the possibilities for radiolabeling the reagent. Figure 5.2 (Chong and Hodges, 1981) shows a reagent possessing several desirable features.

A discussion of the properties of the chemically reactive groups that have been incorporated into bifunctional reagents is beyond the scope of this monograph. Suffice it to say that reagents are available for the modification of all amino acid side chains except for the hydrocarbons, lysine and cysteine being the most popular targets. Many of the reagents are quite selective and react under mild conditions. For details the reader is referred to the references in Table 5.1 and to the following reviews: Means and Feeney, 1971; Glazer et al., 1975; Glazer, 1976; and *Methods in Enzymology, Vols. 11, 25, 46 and 47*.

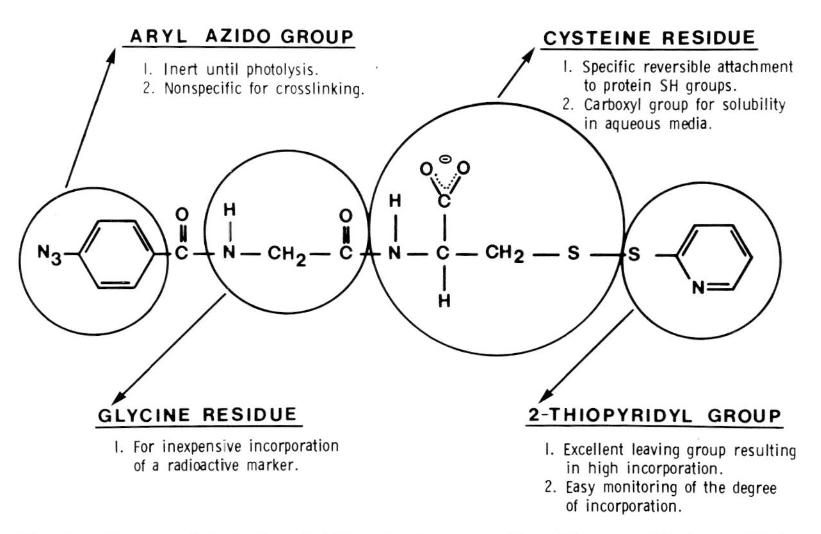

Fig. 5.2. Features of the design of a bifunctional reagent. (From Chong and Hodges, 1981.)

TABLE 5.1

Photochemical, heterobifunctional reagents for crosslinking and for the synthesis of photoaffinity labels

Structure	Group specificity/use	Cleavage	Comments/references
1. *Aryl azides* (a) *Imidates*			
$N_3-C_6H_4-C(=NH)OCH_3$	$-NH_2$	The amidine linkage can be cleaved with 2.15 M CH_3NH_2, pH 11.5, in 75% (v/v) acetonitrile, at 37°C, for 3 h (Packman and Perham, 1982)	1–7
$N_3-C_6H_4-CH_2-C(=NH)OCH_3$	$-NH_2$	The amidine linkage can be cleaved with 2.15 M CH_3NH_2, pH 11.5, in 75% (v/v) acetonitrile, at 37°C, for 3 h (Packman and Perham, 1982)	4
$N_3-C_6H_3(CHO)-O(CH_2)_3-C(=NH)OCH_3$	$-NH_2$	The amidine linkage can be cleaved with 2.15 M CH_3NH_2, pH 11.5, in 75% (v/v) acetonitrile, at 37°C, for 3 h (Packman and Perham, 1982)	8: The aldehyde group may be reduced with NaB^3H_4 to introduce radioactivity
$N_3-C_6H_3(NO_2)-C(=O)NHCH_2-C(=NH)OCH_3$	$-NH_2$	The amidine linkage can be cleaved with 2.15 M CH_3NH_2, pH 11.5, in 75% (v/v) acetonitrile, at 37°C, for 3 h (Packman	9,10 (^{14}C derivative), 11,12

$N_3-C_6H_4-S-S-(CH_2)_2-C(=NH)OCH_3$	—NH_2	Disulfide linkages are generally reduced with thiols such as β-mercaptoethanol. Tributyl phosphine has also been used, e.g. Brunner and Richards (1980)	2,13,14 9
$N_3-C_6H_4-S-S-(CH_2)_3-C(=NH)OC_2H_5$	—NH_2	Disulfide linkages are generally reduced with thiols such as β-mercaptoethanol. Tributyl phosphine has also been used, e.g. Brunner and Richards (1980)	
$N_3-C_6H_4-CH(OH)-C(=NH)OCH_3$	—NH_2	Periodate	4
$N_3-C_6H_4-C(OH)(CH_3)-CH(OH)CH_2-C(=NH)OCH_3$	—NH_2	Periodate	4
(b) N-*Hydroxysuccinimide esters*			
$N_3-C_6H_4-COO-N$ (succinimide)	—NH_2: mild base treatment may be required to cleave the more labile linkages such as esters (Henderson et al., 1978)	–	2,16–19

TABLE 5.1 *continued*

Structure	Group specificity/use	Cleavage	Comments/references
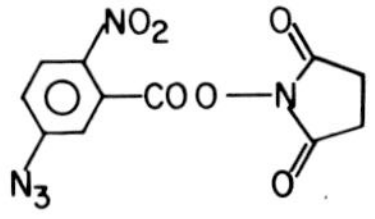 NO_2, COO—N, N_3	—NH_2	–	9–11,16,20
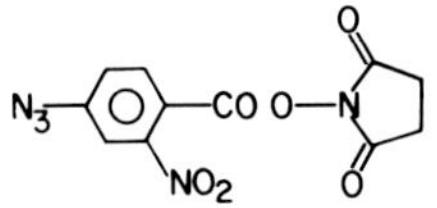 N_3, CO O—N, NO_2	—NH_2	–	21
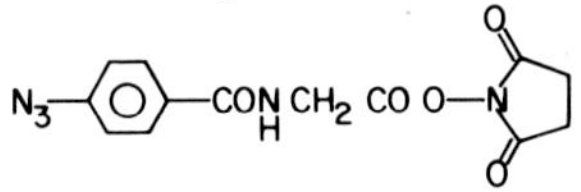 N_3—CONH CH_2 CO O—N	—NH_2	–	16: made with radiolabeled glycine
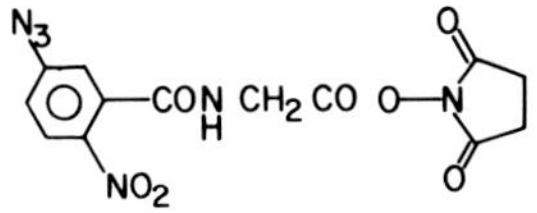 N_3, CONH CH_2 CO O—N, NO_2	—NH_2	–	16,22: made with radiolabeled glycine
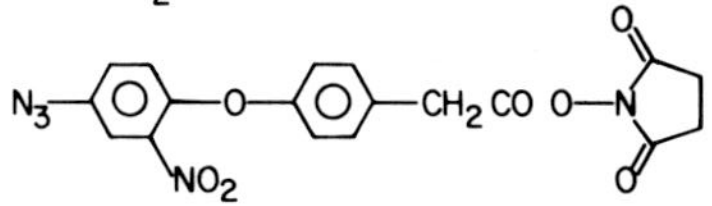 N_3, O, CH_2 CO O—N, NO_2	—NH_2	–	23
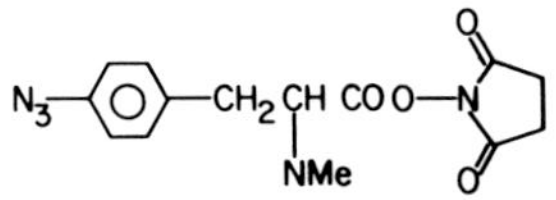 N_3—CH_2CH(NMe) CO O—N	—NH_2	–	4: dimethylamino group improves solubility
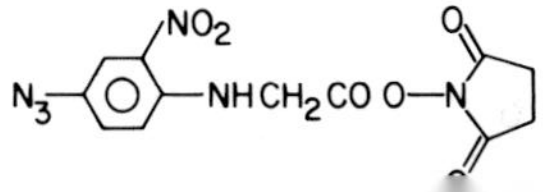 NO_2, N_3—NH CH_2 CO O—N	—NH_2	–	24,25: labeled glycine could be incorporated

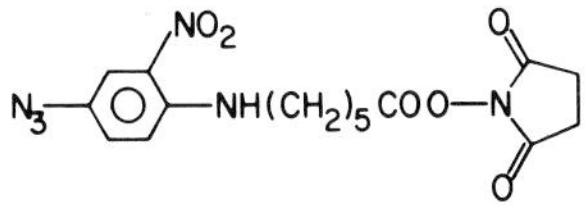 N_3–C$_6$H$_3$(NO_2)–NH$(CH_2)_5$COO–N	—NH_2	–	Based on work in ref. 26. Available from Pierce
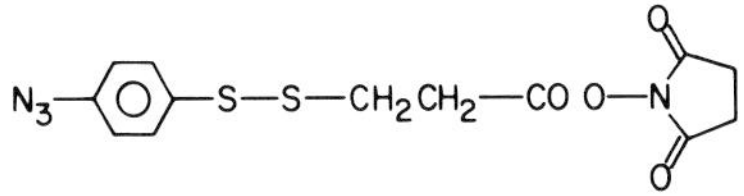 N_3–C$_6$H$_4$–S—S—CH_2CH_2—CO O—N	—NH_2	Thiols	Based on work in ref. 26. Available from Pierce
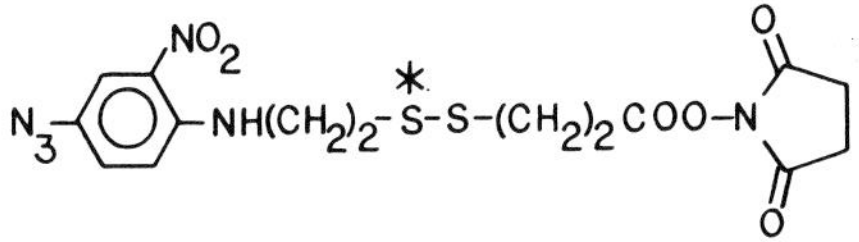 N_3–C$_6$H$_3$(NO_2)–NH$(CH_2)_2$–S*–S–$(CH_2)_2$COO–N	—NH_2	Thiols	26a. Synthesized with ^{35}S at*
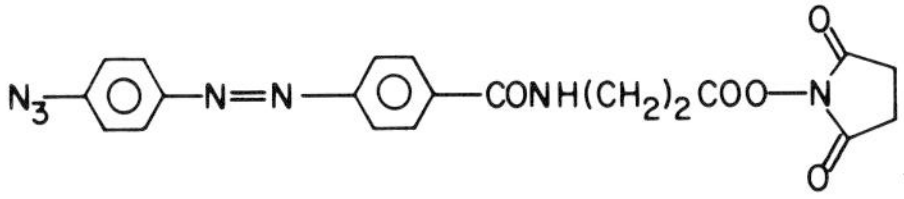 N_3–C$_6$H$_4$–N=N–C$_6$H$_4$–CONH$(CH_2)_2$COO–N	—NH_2	Dithionite	27,28
(c) *Fluoronitroazidobenzenes*			
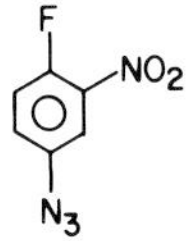 F, NO_2, N_3	Reactive towards nucleophiles. Particularly —NH_2 and —SH	–	29–32
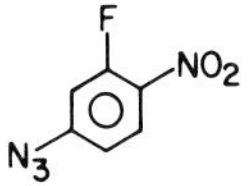 F, NO_2, N_3	Reactive towards nucleophiles. Particularly —NH_2 and —SH	–	33
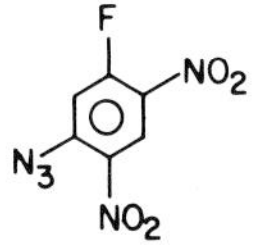 F, NO_2, N_3, NO_2	Reactive towards nucleophiles. Particularly —NH_2 and —SH	Can probably be cleaved from —SH groups at high pH or by RSH	34,35: More reactive than mononitro compound, but —NO_2 group is *ortho* to azide (see Ch. 3)

TABLE 5.1 *continued*

Structure	Group specificity/use	Cleavage	Comments/references
(d) p-*Nitrophenyl ester*			
N_3–C₆H₃(NO_2)–CH_2 CO O–C₆H₄–NO_2	—NH_2	–	25
(e) *Bromoacetyl compounds*			
N_3–C₆H₄–CO–CH_2 Br	—SH	–	36–38: Can first modify —NH_2 with $HS(CH_2)_3$ $C{=}NH(OCH_3)$ and then use this reagent. Available from Fluka and Pierce
N_3–C₆H₄–NHCO CH_2 Br	—SH	–	39
N_3, NO_2–C₆H₃–$COCH_2Br$	—SH	–	40
N_3, NO_2–C₆H₃–$COCH_2$ Br	—SH	–	41
N_3–C₆H₄–CH_2—CH($COOCH_3$)($NHCOCH_2$ Br)	—SH	–	42

(f) *Mixed disulfides*			
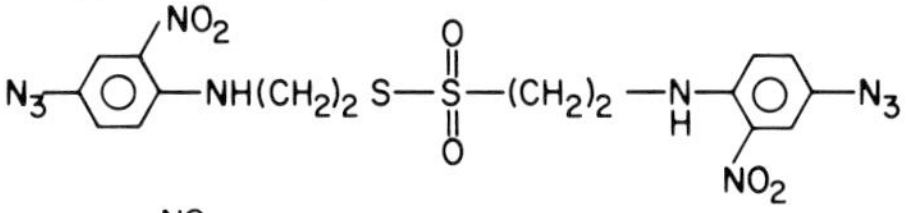	—SH	Thiols	43
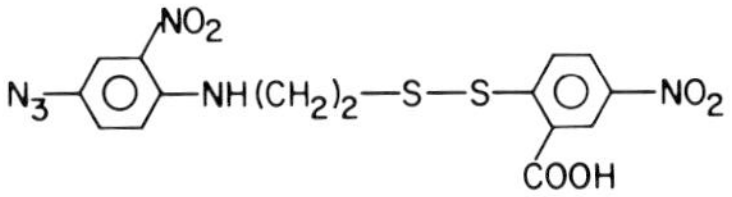	—SH	Thiols	44,45
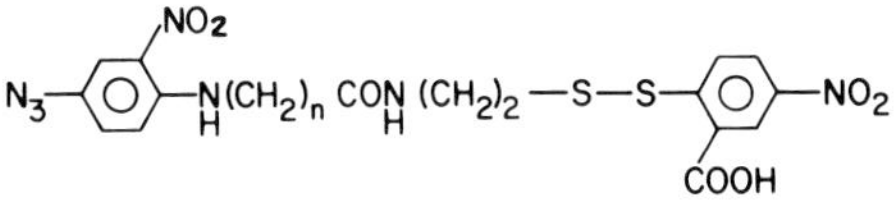	—SH	Thiols	45
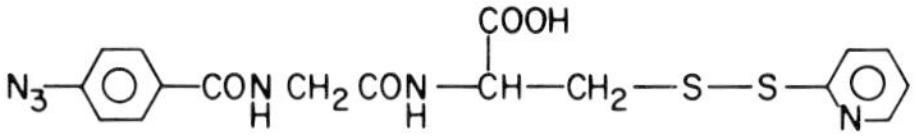	—SH	Thiols	46: Can be made with labeled glycine. See Fig. 5.1
(g) S-*Phthalimido derivative*			
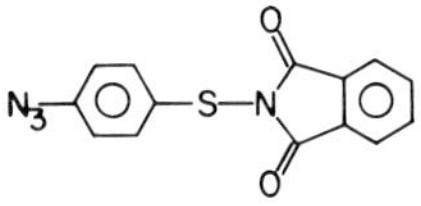	—SH	Thiols	15,15a: Also useful for making mixed disulfides
(h) *Sulfenyl chlorides*			
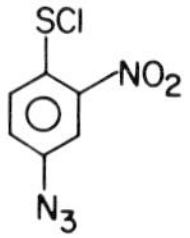	—SH; Trp in acidic media; $—NH_2$ in alkali	By thiols e.g. Muramoto and Ramachandran (1981b)	47-49: 2,4-dinitro-5-azidophenyl sulfenyl chloride is unstable (see 50)
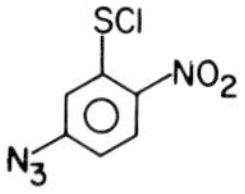	—SH; Trp in acidic media; $—NH_2$ in alkali	By thiols e.g. Muramoto and Ramachandran (1981b)	47-49: 2,4-dinitro-5-azidophenyl sulfenyl chloride is unstable (see 50)

TABLE 5.1 *continued*

Structure	Group specificity/use	Cleavage	Comments/references
(i) *Diazonium salt*			
$N_3-C_6H_4-N_2^+$	His, Tyr, Trp, Cys, Arg, amino groups	In some cases by dithionite (azo linkage)	51
(j) *Dialdehyde*			
$N_3-C_6H_4-CO-CHO$	Arg and to a lesser extent Cys and His. Guanosine	Linkage to Arg cleaved 80% after 48 h, 37°C, pH 7	52,53: Reaction with Arg is minimized by using borate buffer at neutral pH
(k) *Hydrazides*			
$N_3-C_6H_3(NO_2)-CONHNH_2$	Aldehydes, ketones; including dialdehydes formed by oxidizing *cis* diols in sugars with periodate	–	54
$N_3-C_6H_3(NO_2)-NHCH_2CONHNH_2$	Aldehydes, ketones; including dialdehydes formed by oxidizing *cis* diols in sugars with periodate	–	55
(l) *Amines*			
$N_3-C_6H_4-CH_2NH_2$	—COOH, —OPO_3H_2. Attachment using a coupling reagent	–	54
$N_3-C_6H_3(NO_2)-NH(CH_2)_2NH_2$	—COOH, —OPO_3H_2 *(see above). Also trans-glutaminase substrate for attachment to Gln*	–	55–59

N_3–C_6H_3(NO_2)–NH$(CH_2)_2NH_2$	—COOH, —OPO_3H_2 (see above). Also transglutaminase substrates for attachment to Gln	–	60
N_3–C_6H_3(NO_2)–NH CH_2CH_2—S—S CH_2CH_2—NH_2	—COOH, —OPO_3H_2 (see above). Also transglutaminase substrates for attachment to Gln	Thiols	59
N_3–C_6H_3(NO_2)–NHCH_2CH_2 NH–C(=O)–CH(OH)–CH(OH) CO$NH_2CH_2CH_2$ NH_2	—COOH, OPO_3H_2 (see above). Also transglutaminase substrates for attachment to Gln	Periodate	59
(m) *Carboxylates*			
N_3–C_6H_3(NO_2)–NH $(CH_2)_n$ COOH, n = 1–5, 10, 11	Linkage to —OH or —NH_2 using coupling reagents	–	61
(n) *Phosphate*			
N_3–C_6H_4–O–P(=O)(OH)–OH	Linkage to terminal phosphate groups, e.g. of ADP	–	62
2. *Other photoactivatable groups*			
(a) p-*Nitrophenyl ester*			
CF_3–C(=N_2)–COO–C_6H_4–NO_2	Active sites of proteases	–	63,64: Corresponding acid chloride useful in synthesis of reagents. Made with [^{14}C]phosgene (see Ch. 3)

TABLE 5.1 *continued*

Structure	Group specificity/use	Cleavage	Comments/references
(b) N-*Hydroxysuccinimide esters*			
EtOCO–C(=N_2)–COO–N(succinimide)	—NH_2	–	65
CH_3CO–C_6H_4–COO–N(succinimide)	—NH_2	–	16
CH_3CO–C_6H_4–$CONHCH_2COO$–N(succinimide)	—NH_2	–	16: Radiolabeled glycine may be incorporated
C_6H_5–CO–C_6H_4–CH_2CH_2COO–N(succinimide)	—NH_2	–	66
(c) *Pentachlorophenol ester*			
C_6H_5–CO–C_6H_4–$CONHCH_2COO$–C_6Cl_5	—NH_2	–	16: Radiolabeled glycine may be incorporated
(d) *Sulfhydryl*			
EtOCOC(=N_2)CONH CH(COOH) CH_2 SH	—SH	Thiols	67: RNA photochemically derivatized, cross-link then formed by H_2O_2 oxidation to yield disulfides

(e) *Mixed disulfide*			
$N_2CHCOOCH_2$-C$_6$H$_4$-S—S-(2-pyridyl)	—SH	Thiols	68: Synthesized using [^{14}C]glyoxylate
(f) *Isothiocyanate*			
Fluorescein isothiocyanate	—NH_2, —SH	–	69,70: requires O_2
(g) *Maleimide*			
NO_2-C$_6$H$_3$(OCH$_3$)-O(CH$_2$)$_n$CONHCH$_2$N(maleimide)	—SH	–	71

1: Ji (1977); *2*: Ji et al., (1980); *3*: Sutoh (1980); *4*: Rinke et al. (1980); *5*: Andreasen et al. (1981); *6*: Hinds and Andreasen (1981); *7*: Klein et al. (1981); *8*: Maassen (1979); *9*: Lewis et al. (1977); *10*: Lewis and Allison (1978); *11*: Randolph and Allison (1978); *12*: Nathanson and Hall (1980); *13*: Das et al. (1977); *14*: Das and Fox (1978); *15*: Vanin and Ji (1981); *15a*: Moreland et al. (1982); *16*: Galardy et al. (1974); *17*: Yip et al. (1978); *18*: Yeung et al. (1980); *19*: Johnson et al. (1981); *20*: Beneski and Catterall (1980); *21*: Girshovich et al. (1974); *22*: Stadel et al. (1978); *23*: Hsiung et al. (1974); *24*: Hsiung and Cantor (1974); *25*: Thamm et al. (1980); *26*; Lomant and Fairbanks (1976); *26a*: Schwartz et al. (1982); *27*: Jaffe et al. (1979); *28*: Jaffe et al. (1980); *29*: Fleet et al. (1969); *30*: Fleet et al. (1972); *31*: Levy (1973); *32*: Bisson et al. (1978); *33*: Smith and Knowles (1974); *34*: Wilson et al. (1975); *35*: Erecinska et al. (1975); *36*: Hixson and Hixson (1975); *37*: Erecinska (1977); *38*: Schwartz and Ofengand (1978); *39*: Budker et al. (1974); *40*: Seela (1976); *41*: Seela and Rosemeyer (1977); *42*: Klausner et al. (1978); *43*: Huang and Richards (1977); *44*: Witzemann and Raftery (1978); *45*: Witzemann et al. (1979); *46*: Chong and Hodges (1981); *47*: Muramoto and Ramachandran (1980); *48*: Demoliou and Epand (1980); *49*: Muramoto and Ramachandran; *50*: Canova-Davis and Ramachandran (1980); *51*: Escher et al. (1979); *52*: Politz et al. (1981); *53*: Ngo et al. (1981); *54*: Girshovich et al. (1976); *55*: Chicheportiche et al. (1979); *56*: Das and Fox (1978); *57*: Carney et al. (1979); *58*: Lee et al. (1979); *59*: Gorman and Folk (1980); *60*: Darfler and Marinetti (1977); *61*: Jeng and Guillory (1975); *62*: Maassen and Moller (1974); *63*: Chowdhry et al., (1976); *64*: Gupta et al. (1979); *65*: Bispink and Matthei (1973); *66*: Barta et al. (1975); *67*: Oste et al. (1977); *68*: Henken (1977); *69*: Lepock et al. (1978); *70*: de Luca et al. (1981); *71*: Jelenc et al. (1978).

In brief, the chemical arm should react reasonably rapidly with functional groups in the preparation under investigation, at pH 7 to 8 where possible, and at ~ 37 °C or below. To avoid significant perturbations of the organization of the components of the sample by the high concentrations of reagent that are usually required for successful crosslinking, attention must be paid to the nature of the chemical modification. Particular attention should be given to the conservation of charged groups. For example, where amino groups are to be modified imidates are currently favored as the charges on the protonated —NH_2 groups of the lysine residues are preserved as postively charged amidines (Fig. 5.3). Acylation with say an acyl azide or a *N*-hydroxysuccinimide ester would eliminate the positive charge. Related problems can arise with other functional groups, including arginine (modification with reagents related to glyoxal) and aspartic or glutamic acid (attachment of amines using carbodiimides).

Optimal conditions for the chemical modification step can usually be found by comparing the final outcomes of crosslinking experiments with different initial crosslinker concentrations and (or) with different incubation times for the chemical attachment step. Alternatively, a reagent with a chromophoric leaving group may be used in the first step. Two heterobifunctional reagents containing activated disulfides for attachment to sulfhydryl groups in proteins have been equipped in this way. One incorporates the 2-nitro-4-mercaptobenzoic acid leaving group of Ellman's reagent (Witzemann et al., 1979; and Table 5.1.1f) and the other the 2-thiopyridyl leaving group (Chong and Hodges, 1981; Fig. 5.2).

The properties of the photochemical groups found in bifunctional reagents have been described earlier and we have already mentioned that highly reactive photogenerated species would be the most desirable both for obtaining indiscriminate attachment to neighboring molecules, and for distinguishing between permanent and collision complexes. Another im-

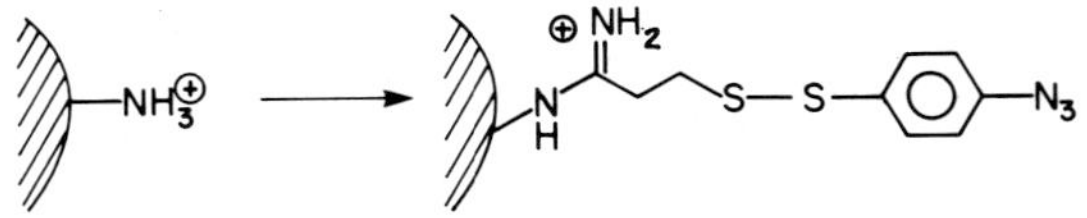

$$\text{—}NH_3^{\oplus} \longrightarrow \text{—}NH\text{—}C(=\overset{\oplus}{N}H_2)\text{—}CH_2CH_2\text{—}S\text{—}S\text{—}C_6H_4\text{—}N_3$$

Fig. 5.3.

portant consideration is that the photochemical arm of the crosslinking reagent should not cause crosslinking in the dark. This might occur if the photoactivatable group were chemically reactive; occasionally reaction in the dark has been observed with photoaffinity reagents (Chapter 4). A second possibility is that a reagent directed at sulfhydryl groups could cause the formation of disulfide bridges (Fig. 5.4). An intramolecular crosslink was formed in low yield in the dark when rabbit muscle aldolase was prepared for photochemical crosslinking with the reagent di-*N*-(2-nitro-4-azidophenyl)cystamine-*S*,*S*-dioxide (Huang and Richards, 1977; Table 5.1.1f). At the time it was thought that a disulfide interchange reaction had occurred but another possibility is that vicinal sulfhydryl oxidation by reaction with the aryl azide took place (Section 3.2.1).

In a number of cases the solubility properties of photochemical crosslinking reagents were not carefully considered before their synthesis. Often an undesirable amount of a water-miscible organic solvent is required to dissolve and transfer the reagent to the reaction mixture. Prospective users of a reagent listed in Table 5.1 should consult the original paper describing the use of the reagent to ensure that the system they are labeling is unchanged by the solvent required. The hydrophobicity of the reagent may also determine the site at which it reacts. This will be an important consideration when membranes are being crosslinked but it may also be a factor with other macromolecular complexes. Several authors have made successful attempts to produce highly soluble reagents by incorporating ionizable groups such as carboxyl or dimethylamino groups into otherwise insoluble molecules (e.g. Chong and Hodges, 1981; Rinke et al., 1980). Reagents containing charged reactive groups such as imidates are usually adequately soluble in buffer (e.g. Rinke et al., 1980). Staros (1982) has recently made water soluble *N*-hydroxysulfosuccinimide esters which might be incorporated in photochemical reagents.

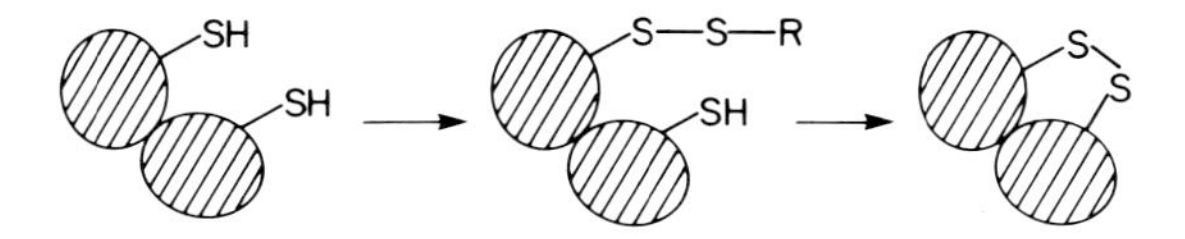

Fig. 5.4.

The permeance of a reagent will be an important factor when attempts are made to crosslink membrane proteins on the external surface of sealed vesicles. To ensure impermeance towards membranes a molecule should possess a charged group, or better two or more charged groups, which ionise at pH values well away from the neutral region. It is probable that none of the photochemical reagents produced so far can be trusted to be membrane impermeant. Indeed, truly impermeant chemical crosslinking reagents, processing two sulfonate groups, have been evaluated only very recently (Staros et al., 1981; Staros, 1982). In designing impermeant or highly soluble crosslinking reagents care must also be taken to avoid amphipathic species that might have detergent-like properties.

A further point that must be considered in the choice or design of a crosslinking reagent is the nature of the cleavable group connecting the two arms. The most popular choice has been the disulfide bond which is easily cleaved by incubation with an excess of a reduced thiol such as β-mercaptoethanol or dithiothreitol. Disulfide interchange may lead to difficulties including crosslinking in the dark or attachment of the reagent to a sulfhydryl group rather than the target sought by the chemically reactive group of the reagent. To minimize interchange during work-up samples are usually treated with a sulfhydryl reagent, e.g. *N*-ethylmaleimide, after the first step of chemical attachment of the reagent (e.g. Markwell and Fox, 1980) or after both chemical attachment and irradiation but before electrophoretic analysis (e.g. Huang and Richards, 1977).

Cleavable linkages other than disulfide bonds have been incorporated into photochemical crosslinking reagents. They include the azo group (—N=N—, cleavable by reduction with dithionite: Jaffe et al., 1980) and the 1,2-diol linkage (cleavable with periodate: Rinke et al., 1980; Gorman and Folk, 1980). The problem of disulfide interchange does not arise with these groups and in the future they may be used more extensively than they are at present.

A final consideration in the design of a photochemical crosslinking reagent is the question of incorporating radioisotopes. Almost all the crosslinking experiments referred to here have been done with unlabeled reagents but there would seem to be distinct advantages to using labeled crosslinkers of high specific radioactivity. Such reagents could be used at low concentrations permitting measurable crosslinking without disturbing a macromolecular assembly by extensive chemical modification. The crosslinking pattern would be clearer as only crosslinked and monofunctionalized species would be visible in autoradiograms. The most straightforward approach to radiolabeling would be to label both arms of the molecule at equal specific radioactivity and with the same isotope. This could be done by separately preparing reagent labeled in the chemically reactive arm and reagent labeled in the photoactivatable arm, and mixing the appropriate amounts of each. Here, it must be ensured that the bonds formed to the crosslinked proteins are stable to the crosslink cleavage conditions. F.M. Richards has suggested the attractive possibility of using unlabeled crosslinkers and cleaving them with a radioactive reagent (e.g. disulfides with [^{35}S]sulfite or diols with periodate followed by [^{3}H]sodium borohydride). A single labeled cleavage reagent would serve several crosslinkers. An obvious drawback is the presence of disulfides and sugars on many polypeptides but the general idea deserves consideration.

5.4. *Experimental procedures*

A photochemical crosslinking experiment usually proceeds as follows. The crosslinking reagent is mixed with the macromolecular assembly that is under investigation and reaction with the chemical arm takes place. Excess reagent is sometimes quenched with a low molecular weight nucleophile and then removed from the sample. If a disulfide-containing reagent is used free SH— groups may be blocked at this stage. Finally, the photochemical crosslinking step is carried out and the crosslinking pattern is analyzed.

No simple guide can be given to the concentration of reagent that must be used in the first, chemical attachment step. Huang and Richards (1977) made a careful study of the extent of modification of sulfhydryl groups in rabbit muscle aldolase and in human erythrocyte membranes with di-*N*-(2-nitro-4-azidophenyl)cystamine-*S*,*S*-dioxide (Table 5.1.1f). Only a slight excess of reagent was required to obtain extensive chemical modification, and there was little advantage in using the reagent at above ~ 50 μg ml^{-1} (0.1 mM) to crosslink ghosts at a concentration of ~ 1 mg (protein) ml^{-1}.

Ji et al. (1980) found that the modification of erythrocyte ghosts with the *N*-hydroxysuccinimide ester of 4-azidobenzoic acid (Table 5.1b) at 0.5 mM led to extensive crosslinking of spectrin on subsequent irradiation. Higher concentrations (2.5 mM) of methyl-3-(4-azidophenyldithio)propionimidate (Table 5.1a) were required to produce a similar outcome which is not surprising as hydrolysis competes strongly with the reaction of imidates with amino groups. Other workers have also found it necessary to use high concentrations of imido ester reagents to obtain extensive crosslinking (Rinke et al., 1980; Markwell and Fox, 1980).

Besides the reagent concentration, the time of reaction, the pH, temperature and the presence of organic solvents are variables that require consideration. Low concentrations of water-miscible organic solvents originating from the addition of the reagent are usually of little consequence. Even pyridine (final concentration 0.5% w/v) has been used with no apparent ill effects on erythrocyte membranes (Huang and Richards, 1977), and alcohols, DMSO, DMF, THF, dioxane or acetonitrile are usually reasonable choices. The investigator should perform experiments to demonstrate that varying the concentration of the organic solvent (from the lowest feasible concentration to a maximum, say ~ 5%) has no effect on the outcome of the crosslinking experiment. An advantage of the photochemical approach is that organic solvents may be removed before the crosslinking step, perhaps reversing any untoward effects.

The appropriate reaction time for the first, chemical, step of crosslinking may often be gauged from the known reactivity of the reagent. Useful information may be obtained from the reviews cited earlier on chemical modification and crosslinking. In the absence of a radioactive reagent or a method for assaying remaining unmodified residues the eventual outcome of the crosslinking experiment may be the only guide to the initial reaction time. Of course it is pointless to extend this period beyond that required for complete hydrolysis of the reagent which occurs in many cases (see, e.g. Peters and Richards, 1977; Ji, 1979). The pH of the medium is often a crucial factor in determining the extent of attachment of a reagent as the ionization states of both the reactive groups and reaction intermediates will determine the rate and course of a modification reaction which is often in competition with hydrolysis or other undesirable events. For example, one of the most useful classes of reagents, the imidates, is one of the most complex in this regard. Reaction at too low a pH value, e.g. 8.0, may lead to hydrolysable imidate bonds rather than the desired amidines. Such bonds may further give rise to undesirable ‘monofunctional’ crosslinks that are resistant to cleavage (Fig. 5.5).

Nevertheless a compromise must be reached between the optimal reaction conditions and the stability of the biological preparation and most crosslinking experiments with imidates have been performed in the range pH 8 to 8.5 with high concentrations of reagents (e.g. Kiehm and Ji, 1977; Ji et al., 1980; Rinke et al., 1980; Markwell and Fox, 1980). Similar dilemmas often arise with other reagents.

Following the chemical attachment step, excess reagent should be removed by dialysis, gel filtration, or for membranes, centrifugation and resuspension. Where possible, some workers take the added precaution of first quenching unreacted reagent (for instance a Tris-containing buffer might be used to stop a reaction with a reagent directed at amino groups), but if the procedure is such that no active reagent remains when the preparation is solubilized for analysis (see below) this step may be omitted. Where disulfide-containing reagents are used the sample may be treated with NEM at this stage, to prevent disulfide interchange. If possible, this step is best deferred until crosslinking by photolysis is completed, as any alteration of the native state of the sample beyond that necessary for crosslinking is undesirable.

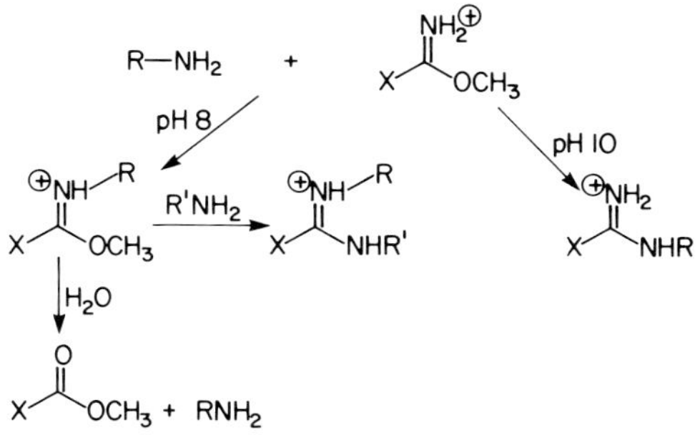

Fig. 5.5.

Most of the principles given for the photolysis step of a photoaffinity labeling experiment (Chapter 4) apply to the photolysis step of a crosslinking experiment. The optimal duration of irradiation can be measured by the extent of crosslinking, perhaps using one-dimensional gel electrophoresis (see below). Crosslinking should not occur on irradiation in the absence of the reagent or when photolysis is omitted but the reagent is present. The sample should be as dilute as possible to prevent random intermolecular crosslinking for soluble proteins or crosslinking between separate membranes, virions, ribosomes, etc. Indeed a change in the crosslinking pattern on dilution can be regarded as diagnostic of intermolecular crosslinking. High protein concentrations may be necessary for extensive chemical modification but the sample should be diluted before irradiation. In most crosslinking experiments surface modification is occurring, and species that might act as scavengers should not be present in the buffer; by analogy with the use of scavengers in photoaffinity labeling experiments it is expected that molecules such as thiols will severely reduce the extent of crosslinking by reacting with photogenerated intermediates. Earlier, I described the electrophilic nature of most photogenerated intermediates used in biological experiments. It follows that the pH of the medium will affect the labeling pattern which will in part reflect the distribution of the most nucleophilic groups in the sample. Such changes must not be ascribed to structural rearrangements without supporting evidence.

The use of flash photolysis (Kiehm and Ji, 1977) cannot, of course, increase the rate of reaction of an *individual* crosslinker after activation and therefore in itself cannot provide a means to distinguish natural and collisional membrane protein complexes. Nevertheless, if the photogenerated intermediates do react rapidly (see Chapter 2), flash photolysis might reduce the extent to which macromolecules reorganize during crosslinking improving the degree to which crosslinking reflects the original macromolecular organization (Ji, 1979).

Two-dimensional gel electrophoresis is the method of choice for analysing the products of crosslinking experiments. The usual one-dimensional electrophoretic techniques are rarely useful except in the simplest cases. Huang and Richards (1977), for example, used slab gels to follow the course of a model reaction in which aldolase was crosslinked with di-*N*-(2-nitro-4-azidophenyl)cystamine-*S*,*S*-dioxide. In this case it was simply a matter of separating monomers, dimers, trimers and tetramers of the single polypeptide chain of the enzyme. In the same study, one-dimensional analysis proved useful in following the time course and the dependence on reagent concentration when erythrocyte membranes were crosslinked. A noteworthy observation made by Huang and Richards was that *intra*mole-

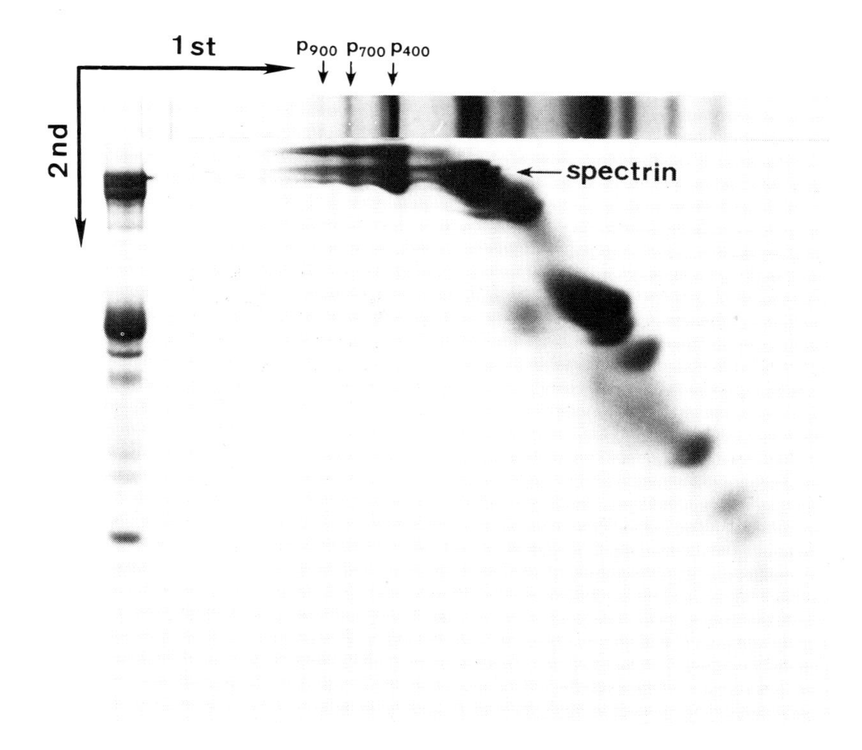

Fig. 5.6. Two-dimensional gel electrophoresis of reversibly crosslinked human erythrocyte membranes. Ghosts were crosslinked with methyl 3-(4-azidophenyldithio)propionimidate and electrophoresed on a 0.5% agarose–1.75% polyacrylamide tube gel. The gel was treated with β-mercaptoethanol (10 mM), and electrophoresed in the second dimension on a 4 to 12% gradient polyacrylamide slab gel. Some of the crosslinked components (mostly oligomers of spectrin) were incompletely cleaved. (Ji et al., 1980).

cular crosslinks may yield polypeptide derivatives that migrate anomalously rapidly in SDS-polyacrylamide gels, compounding the difficulties of a one-dimensional analysis. It is possible to cut bands from a gel, cleave the crosslink and analyse the products on a second gel, but two-dimensional analysis is more convenient.

In two-dimensional gel electrophoresis (Wang and Richards, 1974) the crosslinked sample is first electrophoresed on an SDS-polyacrylamide tube gel which separates the components on the basis of molecular weight. This gel is treated to cleave the crosslinks and transferred to the top of a SDS-polyacrylamide slab gel for electrophoresis in the second dimension. For example, in the case of a crosslinker containing a disulfide bond a layer of agarose containing β-mercaptoethanol may be formed on top of the second gel into which the proteins in the first gel are electrophoresed. After an appropriate delay electrophoresis is continued and crosslinked complexes can be identified from the Coomassie Blue stained polypeptides that lie off the diagonal (Wang and Richards, 1974). The molecular weight of the largest crosslinked species is generally so high that a special agarose-polyacrylamide gel must be used in the first dimension (Wang and Richards, 1974; Kiehm and Ji, 1977). An example is shown in Fig. 5.6.

It will be noticed that in this, and in all other examples of photochemical crosslinking, that crosslinking is far from complete and it seems that a substantial proportion of the photogenerated intermediates must react with buffer components or water, or form highly unstable bonds. Intramolecular crosslinks may be reflected in spots above the diagonal or worse as streaks in the first dimension (Huang and Richards, 1977).

5.5. Possible improvements of existing analytical methods

While the two-dimensional electrophoretic analysis described above is suitable for biological complexes that contain only a few major components (such as the human erythrocyte membrane), it is not feasible to extend its use to more complex systems such as the plasma membranes of many other cell types. Nevertheless, several ways to improve the method may prove useful. Immunoprecipitation of all the crosslinking products derived from a particular polypeptide followed by electrophoresis would undoubtedly be useful where antibodies are available. Similarly if a complex system can be reconstituted with a single radiolabeled component the crosslinks formed to the latter might easily be identified. In such a case, Johnson et al. (1981) were able to identify the glucagon receptor of rat liver membranes by 'photoaffinity crosslinking'. They incubated membranes in the dark with ^{125}I-labeled glucagon and then added the *N*-hydroxysuccinimide ester of *p*-azidobenzoic acid (50 μM). On subsequent irradiation crosslinking, apparently specific, to a 53,000 dalton membrane component occurred. One to two percent of the glucagon became attached to its receptor and at this level of crosslinking no confusing higher molecular weight species were observed. 'Photoaffinity crosslinking' has also been achieved with small organic ligands (Shorr et al., 1982).

The interpretation of two-dimensional gels could also be simplified by using reagents more informative than Coomassie Blue to stain the polypeptides. Markwell and Fox (1980), for example, surface labeled Sendai virus before crosslinking and thus could trace the fate of a subset of the envelope components. The recently introduced Gelcode ™ silver stain (Upjohn Diagnostics) which stains polypeptides in characteristic colors, and Western blots in which polypeptides are transferred from gels to paper or nitrocellulose detected immunochemically are of great potential value (e.g. Burnette, 1981).

5.6. *Examples of photochemical crosslinking*

In a recent instructive example Rinke et al. (1980) used azidoarylimidates to form RNA–protein crosslinks in *E. coli* ribosomes. Five photochemical reagents were tested and only one, methyl-4-azidophenylacetimidate, was found to give a useful extent of crosslinking. Ribosomes (A_{260}: 20 U/ml) were reacted with the imidoester (5 mM) in the dark, at pH 8 to 9, for 30 min at 37 °C. The ribosomes were then precipitated with ethanol and redissolved in buffer at a somewhat lower concentration (5 U/ml) before irradiation. The analysis was greatly aided by the use of biosynthetically labeled RNA or protein. The imidate arm reacted specifically with the polypeptide components of the ribosomes. After irradiation the RNA–protein adducts were isolated by sucrose density gradient centrifugation. Four to eight percent of the total protein was crosslinked to RNA with radiation doses far below those required for crosslinking in the absence of the reagent. After ribonuclease digestion (A and T_1) small fragments of radiolabeled RNA were left attached to the ribosomal polypeptides which were separated by two-dimensional gel electrophoresis. The labeled spots did not exactly correspond to the positions of the Coomassie Blue stained polypeptides and in some cases multiple spots were observed. In a more complex system it would have been extremely difficult to assign the labeled spots to specific polypeptides.

A further experiment was performed that demonstrated the possibilities of the photochemical approach. The reagent was attached in the dark to 30 S ribosomal subunits which were then combined with radiolabeled 50 S subunits to form complete ribosomes. Photolysis failed to reveal extensive intersubunit protein–RNA crosslinking.

In another case in which a number of exemplary control experiments were done, Markwell and Fox (1980) crosslinked the outer membranes of enveloped viruses with methyl 3-[*p*-azidophenyl)dithio]propionimidate. Virus (4 mg protein/ml) was reacted with the imidate (0.1 to 0.5 mM) at 0°C for 30 min at pH 8.5. The reaction was quenched with 50 mM ammonium acetate, 50 mM NEM (30 min, 25°C), and the virus recovered by centrifugation. After irradiation the crosslinked polypeptides were examined in a two-dimensional SDS-polyacrylamide gel. One complication was that the crosslinking pattern had to be compared with a native pattern of disulfide linkages, and a reagent with a different cleavable crosslink may have been a better choice. As mentioned above, the analysis was simplified by the use of surface labeling.

Control experiments showed that all the crosslinks were cleavable. No crosslinks were formed if the reagent was omitted, if photolysis was omitted, or if the reagent was first inactivated by treatment with ammonium acetate. These experiments proved that neither 'monofunctional' crosslinking, nor disulfide interchange to give crosslinking in the dark occurred. If the membranes were solubilized in SDS before irradiation no crosslinking occurred. The results further showed that the reagent (p$K \sim 7$) was membrane permeable.

5.7. *Improved crosslinking reagents*

While the length of Table 5.1 attests to a great deal of activity in the area of photochemical heterobifunctional reagents, there remains room for improvements. Particularly useful would be reagents with photoactivatable groups of the highest reactivity, and reagents with cleavable groups other than disulfide bonds. Four groups of reagents which differ from the run of the mill may inspire the reader to devise yet others. They are the transglutaminase substrates of Gorman and Folk (1980) one of which contains a cleavable diol (Table 5.1.11), the *p*-nitrophenylether-maleimide reagents of Jelenc et al. (1978: Table 5.1.2g), the psoralen derivatives for nucleic acid–protein crosslinking under development in Cantor's laboratory (Cantor, 1980; Schwartz et al., 1983), and the carrier molecules of Aimoto and Richards (1981). The latter are rigid compounds with differing Stokes radii that can be used to transfer photochemical reagents to accessible sites on membranes.

CHAPTER 6

Photoactivatable reagents for studying membrane topography

In recent years there has been considerable interest in the structure of biological membranes, and photochemical reagents capable of yielding low resolution structural information about membrane proteins have been developed. Several examples of photochemical surface-labeling reagents have appeared, and much effort has been devoted to the development of photoactivatable hydrophobic reagents for labeling from within the lipid bilayer.

6.1. Surface labeling reagents

Modification with chemical reagents or enzymes, and digestion with proteolytic enzymes have been used for over a decade in investigations of the topography of membrane proteins. Such studies are capable of revealing which regions of integral and peripheral membrane proteins are exposed in the surrounding buffer. Further, by labeling sealed vesicles, those regions of a protein exposed on one of the two sides of the lipid bilayer may be identified. In favorable cases it is possible to compare the results from the latter experiment with the results obtained by labeling vesicles of the opposite orientation, by labeling membrane fragments (both sides exposed), or by trapping a reagent inside vesicles and labeling from within. The combined data reveal which proteins and which regions of them are exposed on each face of the membrane. Polypeptides or segments of them labeled from both sides must span the bilayer.

Photochemical surface labeling reagents introduced by Staros and Richards (1974) have a number of potential advantages. First, the high reactivity of photogenerated reagents means that many different functional groups can be modified with a single reagent and in favorable circumstances reaction with carbon–hydrogen bonds might occur allowing the labeling of amino acid residues inert towards conventional reagents. Second, it is not an easy matter to trap chemical reagents inside cells, organelles or vesicles as they will react with groups on the outer surface while loading is taking place. Enzymes such as lactoperoxidase may be trapped more readily but the bulkiness of these 'reagents' probably means that some surface residues will be inaccessible to them. Photochemical reagents which may be activated at the will of the investigator are potentially the most useful candidates for labeling from within closed membrane systems.

The first photochemical surface labeling reagent was NAP-taurine (Staros and Richards, 1974) (Fig. 6.1). The negatively charged sulfonate group was intended to confer water-solubility and to make the reagent impermeant to biological membranes. Further desirable properties of surface labeling reagents are that they should not bind tightly to sites on polypeptides, and that the photogenerated intermediates should be highly reactive. In the light of recent experience it seems that the original goal of indiscriminate reaction by a photogenerated intermediate may be extremely difficult to achieve (Staros, 1980; Bayley, 1982; and Chapter 2). It certainly cannot be assumed that the extent of labeling of a membrane protein by existing photochemical surface labeling reagents will be proportional to the exposed surface area of the protein. It should also be recognized that indiscriminancy of reaction and therefore random labeling can only be achieved at the cost of a reduced extent of labeling as extensive reaction with water and buffer components will occur with highly reactive species.

Although NAP-taurine can no longer be considered a generally useful surface labeling reagent (see below), besides developing the general concept, Staros and Richards made many interesting observations in their early

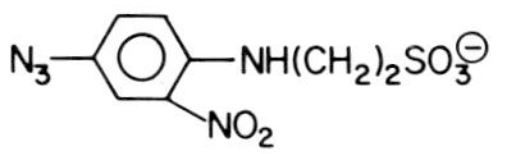

$N_3-C_6H_3(NO_2)-NH(CH_2)_2SO_3^{\ominus}$

Fig. 6.1.

work (Staros and Richards, 1974; Staros et al., 1974; Staros et al. 1975). They labeled high concentrations of intact human red blood cells, or resealed or broken ghosts in a specially designed apparatus. The sample, spread out as a thin film in a cooled rotating horizontal tube, was irradiated with a Sylvania R32 photoflood lamp (Staros and Richards, 1974). NAP-taurine was used at a concentration of 0.1 to 1.9 mM, (^{35}S or ^{3}H at 15 to 500 mCi $mmol^{-1}$) and the photolysis time was optimized to maximize the incorporation of label but minimize cell lysis which was kept below 2%.

It was observed that a number of minor proteins of the erythrocyte membrane (in the band 2.1 to 2.6 region) which cannot be labeled with conventional surface labeling reagents were labeled with NAP-taurine. At the time this was ascribed to the high reactivity of the photogenerated nitrene which allowed it to label exposed polypeptides that did not contain nucleophilic groups. Differences in the labeling pattern of membrane proteins from intact cells and resealed ghosts were detected that were not observed by conventional means. Finally, it was found that the red cell membrane was permeable to NAP-taurine at 37 °C but completely impermeable at 0 °C. This permitted Staros and coworkers to label the cytoplasmic aspect of the membrane in intact erythrocytes.

During the course of these experiments several disconcerting observations were made. Band 3, the anion channel, was by far the most heavily labeled polypeptide. The authors suvgested and it was later confirmed (e.g. Knauf et al. 1978) that NAP-taurine was a photoaffinity reagent for the anion channel. It was also found that certain polypeptides that should have been labeled in broken ghosts (namely bands 1 and 2 [spectrin] and band 5) were not strongly labeled, and it was thought that the reagent might have been repelled by a net charge on these proteins. Indeed, weak interactions can affect reactivity considerably (e.g. Glazer et al., 1975: pp. 123–126), but this observation could also be explained by the chemical specificity of the reagent (Chapters 2 and 3) and its true location with respect to the membrane (see below). Staros and coworkers also observed polymerized material at the origin of their gels and globin multimers. The use of an inert atmosphere might have prevented this crosslinking but it could have been caused by radical reactions triggered by hydrogen atom abstraction by the photogenerated nitrenes.

More recently it has become clear that NAP-taurine passes through other membranes more readily than those of erythrocytes. It has proved difficult, for instance, to prevent its penetration into other types of mammalian cells. This limits its use for labeling membranes from the external medium. An even more important problem is that the molecule appears to bind to cell membranes like a detergent (Dockter, 1979; Richards and Brunner, 1980) and it seems that a number of the minor proteins labeled in intact erythrocytes penetrate the membrane deeply from the cytoplasmic side but do not span it. These proteins are accessible to NAP-taurine molecules that reach down into the hydrocarbon region of the bilayer.

In an application related to the surface labeling of membranes, Matheson and colleagues have labeled ribonuclease with NAP-taurine in an attempt to define which residues are exposed at the surface of the protein (Matheson et al. 1977; Matheson and Scheraga, 1979). When analyzing the results of such experiments the selectivity of the reagent should be taken into account, i.e. it should not be assumed that an exposed leucine residue will be labeled as extensively as say a cysteine residue, but it would be fair to compare the extents of labeling of two different leucines. Further it is clear that long-lived intermediates are formed from azides (for NAP-taurine itself see Mas et al. 1980) and it is possible that transiently exposed residues may be derivatized if they are particularly reactive. The presence of weak binding sites for the charged reagent should also be taken into account.

Dockter (1979) has overcome the major disadvantages of NAP-taurine by synthesizing a new reagent 3-azido-(2,7)-naphthalene disulfonate which does not enter phospholipid vesicles (Fig. 6.2). As one of the charged sulfonate groups is close to the reactive azide it is likely that the latter cannot enter the bilayer even if the reagent does bind to the membrane surface. (In this case the ortho-substituent does not appear to upset unduly the reactivity of the photogenerated species by diverting it into an intramolecular reaction.)

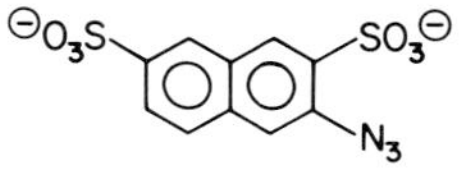

$^{\ominus}O_3S$ $SO_3^{\ominus}$ N_3

Fig. 6.2.

A special feature of the Dockter's reagent is that the reaction products are fluorescent although the sensitivity of detection possible with a radiolabeled reagent cannot be achieved in most laboratories.*

Dockter labeled intact erythrocytes with 3-azido-(2,7)-naphthalene disulfonate at 0.1 to 0.3 mM. At 4 °C, 25 °C and 37 °C, the same pattern of labeling was observed and it closely resembled that seen when intact cells are iodinated with lactoperoxidase. In each case the Coomassie Blue staining pattern of the membrane proteins was unchanged. In erythrocyte ghosts all the major polypeptides were labeled, although some (bands 5, 7 and 8) were only weakly modified which suggests either that they were exposed but were relatively unreactive or that they were much less accessible to the reagent. This reagent appears then to represent an improvement over NAP-taurine. Its synthesis has been described in detail (Moreland and Dockter, 1980).

Rotman and Heldman (1980, 1981) have studied the properties of azidofluorescein diacetate, a reagent which might prove useful for labeling membrane proteins from the inside of cells. The reagent passes through cellular membranes but is hydrolysed inside the cell, presumably by endogenous esterases, to the impermeant azidofluorescein anion. After washing the cells the trapped reagent is photoactivated. Azidofluorescein diacetate has been used to label cytoskeletal proteins but it should be possible to use it or a similar molecule for membrane labeling.

A macromolecular, photochemical surface labeling reagent was developed by Louvard et al. (1976). It comprised a Fab fragment of a human myeloma protein with which 4-fluoro-3-nitrophenylazide (Table 5.1.1c) had been reacted. Used at a concentration of 0.1 mM, this conjugate could be used to label the surfaces of sealed vesicles or it could be trapped inside vesicles and used to label from within. Louvard and colleagues focused their attention on the modification of aminopeptidase in intact brush-border membranes and demonstrated that the enzyme spans the bilayer. From the

* After the reagent is irradiated in the absence of membranes, numerous fluorescent bands are found on SDS-polyacrylamide gel electrophoresis, but they are removed during staining with Coomassie Blue (B. Wisnieski, personal communication). High sensitivity of detection has been achieved using single photon counting techniques for detecting fluorescent bands in gels (M. Dockter, personal communication).

outside, the aminopeptidase in sealed right-side-out vesicles could be labeled to an extent of 7% of the incorporated reagent (5% of the total used). This amounted to 0.25 mol reagent per polypeptide chain. From the inside of the sealed vesicles the extent of labeling was lower as less of the polypeptide chain is exposed to the cytoplasm. The experiment was greatly aided by the availability of an antibody with which the aminopeptidase could be selectively precipitated and a specific proteolytic cleavage that removed the cytoplasmic domain of the protein. The extent of labeling was measured using an ^{125}I-labeled Fab fragment, or at exceedingly high sensitivity with a peroxidase-conjugated anti-Fab immunoglobulin. Some difficulty was experienced in washing the reagent off the outside of re-sealed vesicles in which it had been trapped and it was necessary to account for the residual reagent on the outside of the vesicles in a control experiment in which vesicles were incubated with the Fab conjugate under non-penetrating conditions, thoroughly washed, and then labeled.

If this method is to be extended to membrane preparations for which antibodies against specific components are unavailable, improvements are necessary. The high molecular weight Fab fragments will drastically alter the apparent molecular weights of labeled polypeptides on SDS-polyacrylamide gels making their identification in mixture difficult. Therefore a reversible crosslink is required such that after surface labeling the reagent may be cleaved leaving a small radioactively labeled group attached to the labeled membrane proteins. A number of the molecules listed in Table 5.1 are suitable candidates for incorporation into such a reagent (for a relevant discussion see Schwartz et al., 1982). It should also be noted that the vesicles used by Louvard and colleagues were very large (2000 to 5000 Å in diameter) and had a capacity of ~ 1000 reagent molecules (0.1 mM). Vesicles made by the usual reconstitution procedures have much smaller volumes and will only contain a few molecules of reagent at this concentration.

6.2. *Hydrophobic reagents for membranes*

Photoactivatable, hydrophobic reagents for membranes were developed during the same period as photoactivatable, surface labeling reagents, but because the goals sought with them cannot be achieved by alternative methods they have been more extensively developed.

A hydrophobic reagent is required to label those components of a biological membrane that lie within the hydrocarbon core of the lipid bilayer. The main interest lies in obtaining information about the topography of membrane proteins. First, hydrophobic reagents should label integral and not peripheral membrane proteins. Second, further analysis of a labeled integral protein by peptide mapping permits the determination of which segments of the polypeptide chain lie within the lipid bilayer. Eventually, it is hoped that photochemical reagents suspended at defined points on amphipathic chains will be used to label proteins at defined depths within the lipid bilayer.

The hydrophobic reagents made to date fall into three classes: simple hydrophobic reagents, amphipathic molecules, and phospholipid analogs. In certain particularly favorable cases conventional chemical reagents have been used to label groups within the lipid bilayer (e.g. Sebald et al. 1980; Sigrist and Zahler, 1982; Ross et al. 1982) but on the whole it is expected that the amino acid residues that face the fatty acid side chains or cholesterol molecules of the bilayer will be hydrophobic and chemically inert. Further, the few polar groups within the bilayer are expected to take part in non-covalent intra- or intermolecular bonds and therefore remain inaccessible to hydrophobic reagents. As hydrocarbon residues are the targets, the requirement for an extremely reactive species such as those generated in photochemical reactions is clear. A second important reason for using photoactivatable reagents is that they will not react with surface components while they are equilibrating with the membrane preparation. Hydrophobic reagents have been reviewed (Brunner, 1981; Bayley, 1982; Robson et al. 1982).

6.2.1. Simple hydrophobic reagents

A simple hydrophobic reagent is a small radiolabeled organic molecule that can be photochemically activated after it has bound to a lipid bilayer. The major merit of such reagents in comparison to phospholipid analogs is that they are simple to use with native membranes. Klip and Gitler (1974) introduced 1-azidonaphthalene and 1-azido-4-iodobenzene as hydrophobic reagents. Gitler and his colleagues found that the naphthalene derivative in tritiated form was prone to radiolysis and it also formed 'artefactual polymeric products' when irradiated. 1-Azido-4-iodobenzene has an absorption maximum at 258 nm and was therefore not the ideal choice as a reagent. Some of the problems of the earlier reagents were overcome by using 5-[^{125}I]iodonaphthyl azide (Bercovici et al. 1978). The properties of this reagent and two others that are in current use are given in Table 6.1.

The usual factors involved in the design of photochemical reagents, such as the stability of the molecule and the placement and form of the radiolabel, must also be considered for hydrophobic reagents. Two other features are of particular importance as they determine the performance of the reagents: they are its hydrophobicity, and the reactivity of the photogenerated species. The hydrophobicity of a reagent designed to react within the lipid bilayer should be as great as possible but is constrained by the need to have a molecule of a reasonable size. A relatively hydrophilic reagent may label peripheral proteins and regions of integral membrane proteins exposed to the buffer. This occurs first because a substantial fraction of the reagent is in the aqueous phase at the start of the experiment, and second because a weakly bound reagent may exchange rapidly between the buffer and membrane phases. In the last case a species photogenerated within the lipid bilayer will likely find relatively reactive groups with which to react outside the bilayer. Such phenomena are analogous to the problems of non-specific labeling and pseudophotoaffinity labeling (Section 4.7.3) and may clearly be observed when species of similar reactivity are generated from precursors of varying hydrophobicity. Thus, phenyl azide is a poor reagent, labeling peripheral proteins heavily and reacting efficiently with scavengers in the aqueous phase (Bayley and Knowles, 1978a; 1980). On the other hand, iodonaphthylazide and 1-azido-4-iodobenzene behave well in such tests (Bercovici et al. 1978; Wells and Findlay, 1980; Bayley and Knowles, 1980).

The question of reactivity has two facets. First, a very reactive intermediate will be short-lived and for two reagents of similar hydrophobicity the more reactive one will be less likely to dissociate from and react outside the lipid bilayer (Bayley and Knowles, 1978a,b). Second, as mentioned

TABLE 6.1

Comparison of three well-characterized hydrophobic reagents[a]

Reagent	Structure	$\lambda_{max}/\varepsilon$	P_{rbc}[b]	Reactive species	References
Iodonaphthylazide	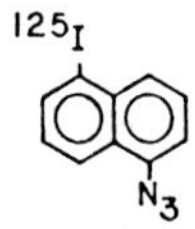 ^{125}I, N_3	310 nm/21,400	163,000	Nitrene and azacycloheptatetraene or benzazirine	Berovici et al. (1978), Kahane and Gitler (1978), Karlish et al. (1977), Tarrab-Hazdai et al. (1980)
Adamantane diazirine	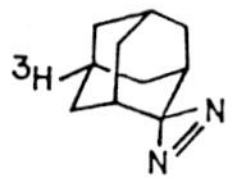 ^{3}H, N, N	372 nm/245	1,750	Carbene and diazoadamantane	Bayley and Knowles (1978a,b, 1980), Goldman et al. (1979), Farley et al. (1980)
3-Trifluoromethyl-3-(*m*-iodophenyl) diazirine	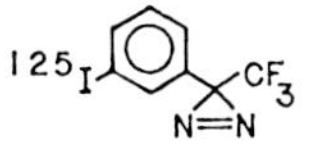 ^{125}I, CF_3, N=N	353 nm/266	24,000	Carbene (diazo compound is inert)	Brunner (1981), Brunner and Semenza (1981), Spiess et al. (1982)

[a] Other simple hydrophobic reagents have been used. For aryl azides see Cerletti and Schatz (1979), Wells and Findlay (1980), Owen et al. (1980), Smith et al. (1981). Pyrene sulfonyl azide has also been used (Sator et al. 1979).

[b] P_{rbc}, the partition coefficient of the reagent into red blood cell membranes, is defined as: (ligand bound/mg membrane protein)/(free ligand/μl buffer).

earlier, many of the residues of polypeptides exposed within the bilayer have hydrocarbon side chains and extremely reactive species such as carbenes are required to derivatize them. For this reason the reagent adamantane diazirine was developed (Bayley and Knowles, 1978b; 1980). The carbenes formed from adamantane diazirine and phenyl diazirine indeed react with saturated fatty acid side chains in phospholipid vesicles (Bayley and Knowles, 1978b), while under comparable conditions phenyl nitrene does not react (Bayley and Knowles, 1978a). However, although adamantylidene is highly reactive and adamantane diazirine ameliorates the second aspect of the reactivity problem, it does not fully solve the first. The problem is that on irradiation adamantane diazirine rearranges, in part, to the linear diazo isomer which is chemically reactive and might dissociate from the membrane to react elsewhere. To prevent the consequent complications, 3-trifluoromethyl-3-(*m*-iodophenyl)diazirine which yields an inert diazo isomer was introduced by Brunner and Semenza (1981).

It should be pointed out that despite their high reactivity carbenes are selective, and while they will react with hydrocarbons in the absence of nucleophiles, a nucleophilic group exposed within the bilayer will preferentially be attacked. Some properties of the three most thoroughly tested reagents iodonaphthyl azide, adamantane diazirine, and 3-trifluoromethyl-3-(*m*-iodophenyl)diazirine are summarized in Table 6.1. There follows a brief description of a typical experimental procedure, a discussion of control experiments, and several caveats, followed by some examples of the applications of hydrophobic reagents.

6.2.2. *Labeling with hydrophobic reagents*

Detailed protocols for labeling membranes with hydrophobic reagents have been described (e.g. Bercovici et al. 1978; Bayley and Knowles, 1980; Brunner and Semenza, 1981). The procedure for labeling erythrocyte ghosts with [^{3}H]adamantane diazirine is typical. Membranes (3 to 4 mg protein ml^{-1}) are sealed under a serum cap in a glass test tube which is then purged with wet N_2. The reagent is then added by syringe in a water-miscible organic solvent, the final concentration of which is kept as low as possible, preferably below 1%. For example 10 μCi of adamantane diazirine (840 mCi $mmol^{-1}$) might be used to label 1 ml of erythrocyte membranes and could be introduced in 10 μl of ethanol. The N_2 stream is discontinued at this stage to avoid entrainment of the quite volatile reagent. An important consideration is that although the average concentration of the reagent is quite low (~ 12 μM), its concentration within the hydrocarbon core of the bilayer is, after equilibration, high (2.1 mM: the concentration of the label remaining in the aqueous phase is 1.4 μM). Some workers, usually lacking a reagent of sufficiently high specific radioactivity, have used millimolar average concentrations of reagents which must result in enormous concentrations within the bilayer. Obviously this is to be discouraged.

After a suitable incubation period (1 h at 4 °C has been used for adamantane diazirine, 5 min at 37 °C for iodonaphthyl azide, and 15 min at 0 °C for 3-trifluoromethyl-3-(*m*-iodophenyl)diazirine), the membrane preparation is irradiated for a period that results in maximal incorporation or a shorter period if membrane damage occurs. After photolysis the membrane preparation is washed thoroughly to remove non-covalently bound photolysis products which can often interfere with subsequent analyses. A useful method is to incubate the membranes with a solution of bovine serum albumin (10 mg ml^{-1}) which binds hydrophobic molecules. After centrifugation and resuspension in fresh albumin-containing buffer the process is continued until no more radiolabel appears in the supernatant. The bovine serum albumin is removed by washing with buffer and the membranes are ready for direct analysis by SDS-gel electrophoresis or for solubilization followed by the isolation of specific polypeptides for analysis by peptide mapping. An alternative method is to precipitate the labeled protein by adding 9 volumes of cold acetone. The precipitate is washed twice with acetone. Wisnieski suggests adsorbing the labeled protein onto silica gel and eluting the non-covalently bound products and the labeled lipids (Hu and Wisnieski, 1979). M13 phage coat protein may be solubilized with buffer containing SDS from the silica at the origin of a TLC plate and then subjected to SDS-polyacrylamide gel electrophoresis.

6.2.3. Control experiments

Several control experiments have been suggested to confirm that hydrophobic reagents do label from within the lipid bilayer (Bercovici et al. 1978; Bayley and Knowles, 1980). Even though the reagents listed in Table 6.1 have been thoroughly tested and their virtues and limitations described, the wise investigator will repeat at least some of the controls described here with his system.

To prove that the peaks of label in an SDS-polyacrylamide gel derive from labeled polypeptides it can be shown that they are still found in preparations that have been extracted with organic solvents (note though that certain proteolipids may be removed) or that they are susceptible to proteolysis. As most hydrophobic peptide fragments remain attached to the lipid bilayer after proteolysis, whereas fragments originating from regions of polypeptides that were exposed to the buffer are often released, proteolysis is a useful test to confirm that labeling occurred in the desired region of the membrane. Proteolysis of membranes solubilized in SDS should alter the electrophoretic mobility of the more recalcitrant polypeptides.

It should next be checked that peripheral proteins are not strongly labeled. This may be done by extracting them selectively (e.g. at low pH: Steck and Yu, 1973) or by measuring the extent of labeling of known peripheral proteins. In certain cases it may be possible to include in the experiment exogenous proteins that bind to the membranes under investigation (Bayley and Knowles, 1980) and antibodies against surface proteins are useful general candidates for this (e.g. Prochaska et al. 1980).

Another useful control is to use a scavenger which resides only in the aqueous phase to mop up any reactive species generated within or migrating into the buffer. Thiols appear to be the most useful scavengers (see Standring and Knowles, 1980; and Section 4.7.4) and glutathione has proved most useful in membrane studies (Bayley and Knowles, 1978a,b; 1980). Molecules that bind relatively weakly to membranes are unsuitable as hydrophobic reagents and will be largely scavenged: a 5-fold reduction in the extent of labeling of erythrocyte membrane proteins by [^{3}H]phenyl azide occurred in the presence of glutathione (16.5 mM). More subtle changes in labeling patterns brought about by scavengers should be interpreted with caution as the scavenger may induce minor changes in the organization of membrane components, mediated perhaps by disulfide bond cleavage when thiols are used.

Control experiments should also be done to ensure that the organic solvent used to introduce the reagent has no effect on the labeling pattern. Concentrations of ethanol of up to 3% (v/v) have been tested without ill-effects on erythrocyte membranes, but other membranes may be more susceptible to damage. Finally, assuming simple partitioning occurs, the ratio of the reagent concentration in the membrane to that in the buffer should be independent of the membrane concentration, and so should the final distribution of label among the polypeptides of the membrane. However, if regions of polypeptides exposed at the membrane surface act as a sink for reactive intermediates in the buffer, these regions will become relatively more heavily labeled at low membrane concentrations (Bayley and Knowles, 1980).

6.2.4. Some caveats for the use of hydrophobic reagents

The first and most important caveat to be considered in interpreting the labeling patterns obtained with hydrophobic reagents is that the reagents do not label polypeptides in an indiscriminate manner. That is to say the extent of labeling of each member of a collection of integral proteins will not be strictly proportional to the surface area of the protein exposed to the hydrocarbon core of the bilayer. It follows that it would be foolish to assume that a polypeptide labeled twice as heavily as another of the same mass is more intimately associated with the lipid bilayer. As peripheral proteins are often lightly labeled by hydrophobic reagents, there may even be borderline cases where a particularly reactive peripheral protein is labeled as heavily as an unusually inert integral protein. Analogous considerations may be applied to the distribution of label within the polypeptide chain of an integral protein. Despite this problem the record of hydrophobic reagents for distinguishing between integral and peripheral proteins, and for delineating which segments of integral proteins lie within the bilayer is good. It must be noted though that several inconsistencies have been found in the results obtained with different reagents used on the same systems, and it seems likely that these inconsistencies must be ascribed to the varied chemical properties of the photogenerated intermediates. Other explanations include the possibility of different initial locations and orientations of the reagents within the lipid bilayer, and different steric and orientation effects on the rates of reaction at different locations caused by the ordered environment of the membrane interior. Several examples of such discrepancies are given in Bayley (1982): the case of Na,K-ATPase is discussed later. The potential user of hydrophobic reagents is advised to use a set of reagents, perhaps those given in Table 6.1, and attempt to obtain a self-consistent set of data.

The region of the hydrocarbon core of the bilayer within which hydrophobic reagents label has not been defined. It is conceivable (Bayley and Knowles, 1978a) that labeling occurs to some extent in the lipid headgroup region where more reactive polar residues are expected to lie. Even if a photogenerated species spent only a small fraction of its lifetime in this reactive zone a substantial amount of labeling could occur there. It can be said with certainty however that for simple amphipathic polypeptides with a single membrane spanning segment, reaction occurs within a region inaccessible to attack by proteases in the intact membranes (e.g. Goldman et al., 1979).

Last, the possibility of labeling hydrophobic pockets in peripheral sites should be considered. Tight, efficiently labeled sites would be derivatized even in the presence of water soluble scavengers. This disadvantage, which does not obtain with photoactivatable lipid analogs, does not appear to have been a severe problem in practice, probably because the concentration of free reagent is usually far below that required to saturate such sites, when they exist. Again, the prudent investigator will use a variety of reagents to reduce the chances of being mislead by such an occurrence, as the site should have some steric specificity. The presence of such binding sites can also be detected by 'prelabeling' the membranes with a high concentration of unlabeled reagent before labeling with the radioactive molecule. In the absence of tight binding sites (or highly reactive residues which may also saturate but at higher reagent concentrations) the labeling pattern should be the same as in the normal labeling experiment (Bayley and Knowles,

1980). A final possibility is to label in the presence of a molecule that bears a strong structural resemblance to the reagent but is not photoactivatable. This is analogous to a protection experiment in photoaffinity labeling.

6.2.5. Some examples of labeling with hydrophobic reagents

In this section the case of glycophorin, which has proved to be straightforward, and the case of Na,K-ATPase, in which some complications have arisen, are briefly discussed.

The erythrocyte glycoprotein, glycophorin A, has provided a useful target on which to test hydrophobic reagents. It has been labeled, in intact erythrocyte membranes, by the nitrene precursors iodonaphthyl azide (Kahane and Gitler, 1978) and iodophenyl azide (Wells and Findlay, 1979), and by the carbene precursors adamantane diazirine (Goldman et al. 1979) and 3-trifluoromethyl-3-(*m*-iodophenyl)diazirine (Brunner and Semenza, 1981). In all four cases the labeled glycophorin was subsequently isolated and then cleaved with trypsin or chymotrypsin to yield a peptide containing the sole membrane-spanning sequence. In each case, this peptide contained most of the label that was incorporated into the glycoprotein, although it is only 10% (w/w) of the intact molecule.

Some peculiarities in the procedures and results are worth noting. About half the iodonaphthylazide incorporated into the proteins of red cell membranes was attached to glycophorin A. This is much more than expected based on the amount of this protein in the membrane and much more than was incorporated in the case of the other reagents.* In the case of 4-iodophenylazide a high concentration of reagent was used in the labeling experiments (1.1 mM overall), higher perhaps than is desirable. When a high concentration of external scavenger was used (50 mM glutathione) the

* The apparent quantitative distribution of label amongst the polypeptides of red cell membranes differs noticeably for the three reagents given in Table 6.1, each of which fulfils the criteria for a useful hydrophobic reagent quite well. Besides the variation in properties between the reagents (Section 6.2.4), practical considerations such as the various electrophoresis systems used (in which some overlaps occur) might partly explain the differences. The three detailed papers on this topic should be read by those intending to use hydrophobic reagents (Bercovici et al. 1978; Bayley and Knowles et al. 1980; Brunner and Semenza, 1981).

extent of labeling of glycophorin A by adamantane diazirine was noticeably reduced (Bayley and Knowles, 1980). Nevertheless, at the level at which it was examined, the distribution of label within the polypeptide was unchanged and peripheral proteins were not strongly labeled in the absence of scavengers. To judge from the results obtained with a photoactivatable phospholipid analog (Ross et al. 1982; and see below), we might suspect that a residue very close to the membrane surface was labeled. It would be of interest to sequence the labeled peptide as the 2-diazoadamantane formed by photochemical rearrangement of the diazirine (Section 3.4.4) might have reacted with one of the glutamic acid residues at the N-terminal end of the hydrophobic sequence. Presumably the unreactive diazo compound formed from trifluoromethyliodophenyldiazirine should not react in this way, but the precise points of attachment of this reagent to glycophorin A also remain to be determined (Brunner and Semenza, 1981).

Although a few finer points of the labeling of glycophorin A with the assortment of reagents quoted above remain to be clarified, the general outcome is clearly satisfactory as are the results of many other experiments with hydrophobic reagents (Bayley, 1982). Occasionally however conflicting data has been obtained emphasizing the need to use several reagents in examining a particular system. One such example is the case of the Na,K-ATPase of eukaryotic cells which contains at least two subunits: the catalytic subunit (α: ~ 100,000 daltons) and a glycoprotein (β: ~ 50,000 daltons).

Both [5-^{125}I]iodonaphthylazide (Karlish et al. 1977) and [^{3}H]adamantane diazirine (Farley et al. 1980) labeled the α-subunit strongly and the β-subunit hardly at all. The simplest interpretation of this result is that the β-subunit is a peripheral protein, tightly attached to the α-subunit. The possibility remains, however, that the β-subunit is relatively unreactive towards hydrophobic reagents even though the two used in these studies have quite different properties. A further possibility that must be taken into account when considering multisubunit membrane proteins is that β penetrates the bilayer but is shielded from the hydrocarbon phase by neighboring α-subunits.*

* Experiments in which the β-subunit was labeled more heavily with amphipathic and phospholipid reagents have been described recently (Montecucco et al. 1981).

When the distribution of label within the α-subunit is considered, the two studies are no longer in agreement. Adamantane diazirine labeled the polypeptide in two broadly defined regions in the COOH-terminal half of the chain (Farley et al., 1980), while iodonaphthylazide labeled regions of the chain both at the COOH-terminus and the NH_2-terminus (Karlish et al. 1977; Jorgensen et al. 1982). It is disconcerting that the azide labels only the NH_2-terminus at low reagent concentrations. It is not clear whether this is at a tight binding site that is saturated before the COOH-terminus becomes derivatized, or whether it is a question of reactivity (see above).

6.3. Amphipathic reagents for membranes

Amphipathic reagents have been used extensively to label biological membranes by Wisnieski and her group. A reagent in current use is 12-(4-azido-2-nitrophenoxy)stearoyl [1-^{14}C]glucosamine which can be made at a specific radioactivity of 50 mCi $mmol^{-1}$ (Fig. 6.3).

Many of the matters that were considered for simple hydrophobic reagents apply to amphipathic molecules and they will not be discussed again here. Like hydrophobic reagents, properly designed amphipathic reagents are convenient to use as they may be added directly to samples

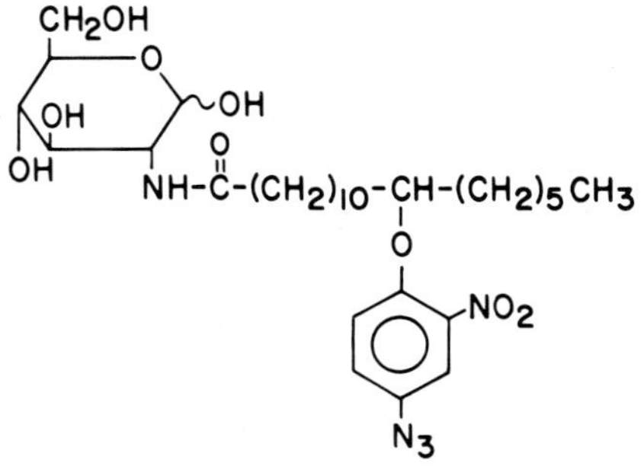

Fig. 6.3.

containing membranes.* In a preparation of sealed membranes or vesicles it is believed that amphipathic reagents, by analogy with similar spin-labeled molecules, are initially restricted to the outer half of the lipid bilayer (Wisnieski and Bramhall, 1981; Wisnieski and Iwata, 1977). Simple hydrophobic reagents, on the other hand, are expected to move freely within the bilayer. This restricted movement of amphipathic reagents potentially allows an extra degree of resolution in mapping the topography of membrane proteins. Further resolution still would be obtained if it could be assumed that the distance between the site of reaction of the photoactivatable group and the position of the hydrophilic headgroup was determined solely by the length of the extended hydrocarbon chain of the molecule. Unfortunately this cannot be assumed to be the case. The site of reaction will be determined by the distribution of conformations available to the reagent during the lifetime of the photogenerated species and the nature of the reactive groups available on the neighboring protein residues in each conformation. Reaction may occur in a less frequently assumed conformation if a particularly reactive group is present there (Bayley and Knowles, 1978a). The frequency of occurrence of undesirable conformations (e.g. a loop such that the photogenerated species lies close to the membrane surface) may be greater with amphipathic reagents, which often have relatively polar groups within the bilayer, than with naturally occurring lipids. Reaction at a depth greater than that determined by the extended chain is presumably less likely than reaction close to the membrane surface.

In designing and using amphipathic reagents the detergent-like nature of the molecules must be considered. It is critical to use a reagent with a specific radioactivity high enough that readily detectable labeling will occur at low reagent concentrations. A systematic study of the effects of reagent concentration on the labeling pattern of a model membrane has not been made, and a prudent approach would be to keep the molar ratio of lipid to reagent above 100. 12-(4-azido-2-nitrophenoxy)stearoyl[1-^{14}C]gluco-

* Wisnieski and colleagues have noted that some of the ^{125}I-labeled amphipaths which they prepared earlier (Iwata et al., 1978) react in the dark with membrane components (personal communication).

samine (Fig. 6.3) itself seems to be a poor detergent (Iwata et al. 1978; and for related compounds Fieser et al., 1956). Another variable which has received little consideration is the nature of the hydrophilic headgroup which might have a profound effect on the lateral distribution of the reagent in the membrane.

In a typical labeling procedure (Wisnieski and Bramhall, 1981), Newcastle disease virus (0.045 mg protein ml^{-1}), the particles of which are bounded by a membrane, were incubated with 12-(4-azido-2-nitrophenoxy)stearoylglucosamine (50 mCi $mmol^{-1}$) at a concentration of 1.3 μM. After 15 min at 37 °C the sample was irradiated.

In an experiment which well illustrates the potential of photogenerated reagents, Wisnieski and Bramhall (1981) labeled cholera toxin during the course of its penetration into the membranes of Newcastle disease virus which had been equilibrated with the glucosamine derivative. At various points during the incubation of toxin and virus at 37 °C a portion of the mixture was irradiated for 15 s which resulted in complete photolysis. Cholera toxin contains disulfide-linked A_1- and A_2-subunits (one of each), and five B-subunits. Over the course of 15 min the extent of labeling of the A_1-subunit of the toxin rose and then fell while the other subunits were unlabeled. It was concluded that only A_1 associates with the bilayer and the remaining protein remains attached at the membrane surface. It was further suggested that after a few minutes A_1 equilibrates between the two halves of the bilayer while remaining attached to A_2 at the surface, and as the amphipathic reagent is present only in the outer monolayer the extent of labeling therefore falls. An equally interesting explanation for the time-dependent reduction in the labeling of A_1 is a conformational change in which a particularly reactive group is obscured or in which a protein of the virus envelope associates with A_1 and shields it from the hydrocarbon phase.

6.4. Lipid derivatives

Phospholipid derivatives with photoactivatable groups on the fatty acyl chains were first made by Chakrabarti and Khorana (1975). These molecules have been introduced into lipid bilayers and used to label the membrane-associated regions of integral membrane proteins. In principle photoactivatable lipids are used in a similar way to the simple hydrophobic and the amphipathic reagents discussed above. The emphasis in this section will therefore be placed on the differences between the groups of reagents. Phospholipid analogs have the advantage that they more closely resemble natural membrane components, and the disadvantage that they usually do not bind spontaneously to lipid bilayers.

An assortment of photoactivatable groups have been attached to fatty acyl chains including aryl and alkyl azides, diazirines, and diazo groups (Table 6.2). Almost all the lipids in current use contain photoactivatable

TABLE 6.2
Photoactivatable lipids in current use

$R_2C(=O)-O-CH(CH_2OCOR_1)-CH_2-O-P(=O)(O^-)-O(CH_2)_2\overset{+}{N}(CH_3)_3$

	R_2	R_1
a	(3-diazirinylphenyl: CH, N=N)$-O(CH_2)_{10}-$	$CH_3(CH_2)_{14}-$
b	$CF_3-C(=N_2)-C(=O)-O(CH_2)_{11}-$	$CH_3(CH_2)_{14}-$
c	(N=N, CF_3 diazirine)$-C_6H_4-(CH_2)_2-S-S-(CH_2)_7-$	$CH_3(CH_2)_{14}-$
d	$N_3-C_6H_3(NO_2)-NH(CH_2)_{11}-$	$CH_3(CH_2)_{12}$
e	$NO_2-C_6H_3(N_3)-$	$CH_3(CH_2)_{14}-$

groups attached to the fatty acyl chain in the 2-position (Table 6.2). As I argued earlier it seems preferable to use a carbene precursor rather than a nitrene precursor for labeling within the bilayer. Two lipids from which carbenes are generated have been used extensively by Khorana and his colleagues (Table 6.2a,b) and more recently Brunner and Richards (1980) have made the trifluoromethylphenyl diazirine derivative, Table 6.2c. The advantages of the latter class of reagents were discussed earlier. Two arylazido lipids which have been used extensively are also illustrated (Table 6.2d,e): one contains a photoreactive group at the end of a long chain (d), while the second (e), is perhaps more akin to lysolecithin and has a reactive group near the membrane surface. Details of the syntheses and use of photoactivatable lipids may be found in the articles of Radhakrishan et al. (1981); Ross et al. (1982); Robson et al. (1982); Brunner and Richards (1980); Bisson et al. (1979); and Bisson and Montecucco (1981).

Phospholipids with smaller, non-polar photoactivatable groups incorporated into their fatty acyl chains would be an improvement over the reagents (listed in Table 6.2). Alkyl azides have been used extensively by Stoffel and his group (see for example Stoffel et al. 1978). Such reagents are attractive because the small size of the photoactivatable group minimally perturbs the lipid bilayer, and has allowed biosynthetic incorporation into virus and bacteria. However, the use of alkyl azides may be questioned on the grounds that they absorb light only weakly ($\varepsilon \sim 25$) with an absorption maximum close to 290 nm, and that the photogenerated nitrene is expected to rearrange readily to an alkyl imine which presumably hydrolyses rapidly to an aldehyde or ketone in aqueous surroundings (for a recent discussion see Kyba and Abramovitch, 1980). Recently, Stoffel and colleagues have produced surprising evidence for the insertion of alkyl nitrenes generated within a bilayer into neighboring fatty acids (Stoffel et al., 1982; Stoffel and Metz, 1982; and Section 3.2.6), but the problem of inefficient photolysis, at wavelengths that are too short, remains. A noteworthy attempt to decrease the bulkiness of the photoactivated group by using $\alpha,\alpha,\alpha',\alpha'$-tetrafluorodialkyldiazirines failed because intramolecular reactions of the photogenerated carbene predominated (Erni and Khorana, 1980). Further work in this area is necessary.

The fact that membrane-bound lipids cannot exchange into the aqueous phase is both a blessing and a curse. Once a photoactivatable lipid has been introduced into a membrane none of the problems associated with labeling from the aqueous phase can arise, as they may with poorly designed hydrophobic reagents. Conversely, the methods that must be used to introduce the lipids into membranes are not as straightforward. They include the reconstitution of membrane systems from dissociated lipids and proteins (directly or by sonication, freeze-thaw, cholate dialysis, etc.: e.g. see Bisson et al. 1979; Brunner and Richards, 1980; Radhakrishnan et al. 1980), membrane fusion, the use of phospholipid exchange proteins (e.g. Brunner and Richards, 1980; Robson et al. 1982) and the biosynthetic incorporation of fatty acid analogs into the lipids of natural membranes (e.g. Greenberg et al. 1976; Quay et al. 1981). The use of phospholipid exchange proteins offers the possibility of labeling only those polypeptides associated with a single leaflet of the lipid bilayer. Certain lipid derivatives with high critical micelle concentrations partition rapidly into biological membranes when added in the form of phospholipid vesicles, but their dissociation rate is slow enough to permit washing of the membranes before photolysis (J. Brunner, personal communication).

As membrane proteins vary so much in their properties no attempt will be made here to give general procedures for their reconstitution with photolabile lipids. The reader is instead referred to the articles quoted in this section and must adapt the methods described to his own ends. The phospholipids are always diluted with non-reactive lipids and again it is impossible to generalize about the ratio that should be used. Molar dilutions of 1:250 to 1:1000 have been used with the arylazido lipids (Table 6.2d,e) for labeling cytochrome *b* (Bisson et al., 1979), mitochondrial ATPase (Montecucco et al. 1980), cytochrome oxidase (Prochaska et al., 1980) and succinate-ubiquinone reductase (Girdlestone et al., 1981). Much lower dilutions in the range of 1:10 to 1:30 have been used by others (e.g. Brunner and Richards, 1980; Radhakrishnan et al., 1980; Ross et al., 1982). Once formed, the reconstituted systems are irradiated and subsequently worked-up by similar methods to those given for other labeling experiments (Chapter 4; and see below). Particular care must be taken to remove non-covalently bound phospholipids (e.g. Radhakrishnan et al., 1980; Quay et al. 1981; Ross et al. 1982).

Two problems associated with the use of phospholipid reagents deserve discussion. One, proper reconstitution, is of definite importance, while the importance of the second, looping around of the fatty acyl chain, is uncertain. Several cases are known in which membrane proteins that are attached to the lipid bilayer by a single hydrophobic sequence of amino acids can be arranged in a number of ways. Included in this category are cytochrome *b* (Enoch et al. 1979) and the coat protein of M13 phage (Wickner, 1976) both of which can be incorporated into bilayers in a transmembrane and in a hairpin conformation. Structurally more complex membrane proteins may also assemble into membranes in a number of ways and where possible it should be confirmed that the native conformation has been attained by using methods appropriate to the system under study (e.g. activity measurements, physical measurements including circular dichroism, proteolytic cleavage experiments, etc.). Photoactivatable phospholipids in single-bilayer vesicles of small radius of curvature and those in multilamellar systems or vesicles of larger radius behave differently. This is evident from experiments in which lipid–lipid crosslinking products were examined (Gupta et al. 1979; for related experiments using amphipathic probes containing benzophenone groups see Czarniecki and Breslow, 1979). Clearly the organization of the fatty acyl chains in highly curved and in essentially planar bilayers differs and there is no reason to suppose that this difference should not also be found in bilayers containing polypeptides. Further, the organization of membrane-bound segments of polypeptides may also differ in these cases. Therefore it would not be surprising if labeling patterns obtained from membrane proteins varied subtly when different methods for bilayer preparations are used (for instance in the quantitative distribution of label within peptides or in the precise point of attachment within a given peptide). Qualitatively or at low resolution though, the results should be similar and this has been born out in a number of cases.

The looping around of photoactivated fatty acyl chains to the membrane surface, with subsequent reaction there or at an intervening point, is a distinct possibility (Bayley and Knowles, 1978a). Such events would be analogous to the exchange of simple hydrophobic reagents into the aqueous phase resulting in the derivatization of peripheral proteins, but in the case of the phospholipid analogs the outcome would not be so severe, and at worst labeling would occur at the margins of the membrane-spanning peptides, rather than at a point within the bilayer. External scavengers are expected to be ineffective in eliminating labeling of this sort, which is unlikely to result in the labeling of peripheral proteins. It should be clear from the previous discussions of the specificity of photogenerated reagents that reaction at positions other than that determined by the outstretched length of the fatty acyl chain will be promoted by the presence of nucleophilic groups on the polypeptide chains in polar regions such as the membrane–water interphase. It is not possible to extrapolate from the distribution of the points of insertion of a reagent into neighboring saturated acyl chains in phospholipid vesicles (Gupta et al. 1979; Czarniecki and Breslow, 1979; Radhakrishan et al., 1982), to the distribution along a membrane-spanning polypeptide, as in the former, but not the latter, all positions are of equal intrinsic reactivity (for further discussion see: Brunner and Richards, 1980; Bayley, 1982).

Two instructive examples of the use of phospholipid reagents are the labeling of succinate–ubiquinone reductase (complex II) (Girdlestone et al., 1981) and the labeling of glycophorin A (Ross et al., 1982).

Complex II consists of four polypeptide subunits (70,000, 27,000, 13,500, and 7,000 daltons). The two larger subunits are components of succinate dehydrogenase and this enzyme or the entire complex II was directly incorporated into phospholipid vesicles at the high protein lipid ratio of 1 : 2 (w/w). The phospholipid (egg yolk lecithin) contained a small amount of lipid (d) (Table 6.2; 1 molecule in 1000) or lipid (e) (Table 6.2; 1 in 400). After irradiation the labeling pattern was analyzed directly by SDS-polyacrylamide gel electrophoresis. Presumably the lipids that did not become attached to polypeptides ran at the dye front, although this was not demonstrated.

A number of interesting observations were made. When the reconstituted succinate dehydrogenase was labeled in the absence of the smallest subunits the lipid (e) labeled both the 70,000 dalton and the 27,000 dalton subunits, while lipid (d) labeled only the smaller subunit strongly. The authors concluded that the 70,000 dalton subunit was peripheral and the 27,000 dalton subunit integral. The small amount of label from (d) that was incorporated into the large subunit (~ 1/10 of that in the smaller subunit) was attributed to the 'folding back' of the arylazido fatty acid to the membrane surface and appeared to be of minor importance in this case. It was also shown that at temperatures below the phase transition temperature of the major lipid component of the vesicles the smaller subunit (27,000) was no longer heavily labeled by lipid (d) suggesting that it does not associate with the hydrocarbon core of the bilayer at these temperatures. Such an effect has not been observed with other membrane preparations but the possibility should be kept in mind. Another interesting observation was that the 70,000 dalton subunit was not labeled by lipid (e) in intact complex II suggesting that the 13,500 dalton and 7,000 dalton subunits (which were labeled by lipids (d) and (e)) serve to hold the large subunit away from the membrane surface in complex II. It was also noted that the same results were obtained whether vesicles were prepared by direct insertion, sonication or cholate dialysis.

Glycophorin A has been labeled with the diazirino lipid (a) (Table 6.2; Ross et al. 1982). Dimyristoyl phosphatidyl choline, lipid (a), and glycophorin A, in the molar ratio 277:30:1, were assembled into vesicles by the cholate dialysis procedure. After irradiation, the non-covalently attached lipids were separated by gel filtration on Sephadex LH60 in formic acid–ethanol–water (20:50:30) leaving 1.8% of the original lipid (a) attached to the glycoprotein. Proteolytic cleavage followed by Edman degradation of the major labeled fragment revealed that Glu-70 was the main site of attachment. This residue also reacted with the hydrophobic chemical reagent dicyclohexylcarbodiimide. When the covalent peptide–lipid adducts were treated with dilute base they were largely cleaved, confirming that ester linkages had been formed by the reaction of a carbene or diazo intermediate with a carboxylic acid. It was also observed that most of the labeling took place in the dark after photolysis with a $T_{1/2}$ of ~ 15 min suggesting that the diazo rearrangement product was predominantly involved. This is consistent with the selectivity for a glutamic acid residue, and the surprising observation that the segment of glycophorin in the second leaflet of the bilayer was unlabeled.

Glu-70 is at the N-terminal border of the membrane-spanning α-helical segment of glycophorin and had previously been placed, because of its polar nature, just outside the bilayer. This new evidence shows that Glu-70 is likely to lie within the lipid bilayer but as we saw earlier it might be closer to the surface of the membrane than the length of the derivatized fatty acyl chain at first suggests, because the reactive intermediate formed from the diazirine could migrate to and react in the membrane-buffer interphase. Indeed, when the diazirine was placed on a chain five carbons shorter than that of lipid (a), the same result was obtained. It is interesting that Glu-72 was unlabeled by the lipid or the carbodiimide, although the above would suggest that it is in the bilayer. It may be protected by a non-covalent bonding interaction with another amino acid residue in the membrane, perhaps within a glycophorin A dimer as suggested in the model of Ross et al. (1982).

In summary, photochemical reagents are available that can be used to distinguish between integral and peripheral membrane proteins. These reagents will label only those segments of integral proteins associated with the bilayer. Several caveats must be noted that arise from the specificity of the photogenerated intermediates for the more reactive amino acids. In particular it appears that these reagents might label reactive groups in the headgroup region of the bilayer and that it may not yet be possible to label polypeptide chains at a defined depth within the bilayer, a goal which might be obtained with more rigid reagents.

List of suppliers

This list has been divided into two parts: (i) suppliers of photochemical equipment and (ii) suppliers of photochemical reagents. The readers' attention is also drawn to five useful directories which can be found in most libraries.

(a) The *Analytical Chemistry* annual Lab Guide (e.g. the 1983–1984 Guide was Analyt. Chem. *55*(No. 10 August) 1–276 (1983).
(b) The *Nature Directory of Biologicals: 1983 Buyers Guide* published by Macmillan Journals.
(c) *Review of Scientific Instruments,* American Association for the Advancement of Science, Washington, D.C.
(d) *Chem Sources – USA:* Directories Publishing Co. Inc., P.O. Box 1372, Ormond Beach, FL 32074.
(e) *The Chemists Companion.* A.J. Gordon and R.A. Ford, Wiley-Interscience, New York, 1972.

(i) *Suppliers of photochemical equipment*

Ace Glass Inc. 1430 N. West Blvd. Vineland, NJ 08360 U.S.A. (609)692-3333	Lamps and glassware for large-scale photolyses
Corning Glass Works Houghton Pk., Corning, NY 14830 U.S.A. (607)974-9000	Filters

Corning Ltd. 1 Princes St. Richmond Surry, U.K. 01-948-2137	
E. Leitz (Instruments) Ltd. 48 Park St. Luton LU1 3HP U.K. 0582-413811	Optical components, including lenses and filters
Melles Griot 1770 Kettering St. Irvine, CA 92714 U.S.A. (714)556-8200	Optical components, including lenses and filters
Oriel Corporation 15 Market Street Stamford, CT 06902 U.S.A. (202)357-1600	All photochemical equipment including lamps, filters, radiation measuring devices
Southern New England Ultraviolet Co. 55 Connolly Pkwy. Hamden, CT 06514 U.S.A. (203)248-2352	Rayonet lamps, quartz vessels
Ultra-Violet Products Inc. 5100 Walnut Grove Ave. San Gabriel, CA 91778 U.S.A. (213)285-3123	Mineralight lamps
UVP Ltd. Science Park Milton Road Cambridge CB4 4BN U.K. 0223-355-722	

(ii) *Suppliers of photochemical reagents*

Amersham Corp.
2636 S. Clearbrook Dr.
Arlington Heights, IL 60005
U.S.A.
(800)323-6695

Radiolabeled 8-azido purine phosphates, reagents for membranes, heterobifunctional reagents

Amersham International
White Lion Rd.
Amersham
Bucks HP7 9LL
U.K.
02404-4444

Molecular Probes, Inc.
24750 Lawrence Rd.
Junction City, OR 97448
U.S.A.
(503)998-6254

Aryl azides: fluorescent hydrophobic and surface labeling reagents

New England Nuclear
549 Albany St.
Boston, MA 02118
U.S.A.
(617)482-9595

Several radiolabeled photoaffinity reagents including 8-azido adenosine phosphates. Radiolabeled heterobifunctional reagents

New England Nuclear
Postfach 401240
D-6072 Dreieich
F.R.G.
06103-85034

Pierce Chemical Co.
PO BOX 117
Rockford, IL 61105
U.S.A.
(815)968-0747

Aryl azides including numerous heterobifunctional reagents. *p*-Nitrophenyl 2-diazo-3,3,3-trifluoropropionate

Pierce Chemical Co. 44 Upper Northgate St. Chester Cheshire CH1 4EF U.K. 0244-382525	
Schwartz-Mann Inc. 2 Ram Ridge Rd. Spring Valley, NY 10977 U.S.A. (800)431-2800	Radiolabeled 8-azidoadenosine and guanosine phosphates
Tridom Chemical Inc. 255 Oser Avenue Hauppage, NY 11787 U.S.A. (516)273-0110	4-Azidoaniline, 4-azidophenacyl bromide

References

ABERCROMBIE, D.M., MCCORMICK, W.M. and CHAIKEN, I.M. (1982) J. Biol. Chem. *257,* 2274.

ABRAMOVITCH, R.A., HOLCOMB, W.D. and WAKE, S. (1981) J. Am. Chem. Soc. *103,* 1525.

AIBA, H. and KRAKOW, J.S. (1980) Biochemistry *19,* 1857.

AMITAI, G., AVISSAR, S., BALDERMAN, D. and SOKOLOVSKY, M. (1982) Proc. Natl. Acad. Sci. U.S.A. *79,* 243.

ANDREASEN, T.J., KELLER, C.H., LAPORTE, D.C., EDELMAN, A.M. and STORM, D.R. (1981) Proc. Natl. Acad. Sci. U.S.A. *78,* 2782.

ATHERTON, J.H. and FIELDS, R. (1968) J. Chem. Soc. (C) 2276.

BAKER, B.R. (1964) J. Pharm. Sci. *53,* 347.

BARTA, A., KUECHLER, E., BRANLANT, C., SRI WIDADA, J., KROL, A. and EBEL, J.P. (1975) FEBS Lett *56,* 170.

BARTLETT, P.A. and LONG, K.P. (1977) J. Am. Chem. Soc. *99,* 1267.

BASTOS, R. DE N. (1975) J. Biol. Chem. *250,* 7739.

BAYLEY, H. (1979) Z. Naturforsch. *34c,* 490.

BAYLEY, H. (1982) In: Membranes and Transport, Vol. 1, A. Martonosi, ed., Plenum, New York, pp. 185–194.

BAYLEY, H. and KNOWLES, J.R. (1977) Methods Enzymol. *46,* 69.

BAYLEY, H. and KNOWLES, J.R. (1978a) Biochemistry *17,* 2414.

BAYLEY, H. and KNOWLES, J.R. (1978b) Biochemistry *17,* 2420.

BAYLEY, H. and KNOWLES, J.R. (1980) Biochemistry *19,* 3883.

BAYLEY, H., STANDRING, D.N. and KNOWLES, J.R. (1978) Tetrahedron Lett. p. 3633.

BAYLEY, H., HUANG, K.S., RADHAKRISHRAN, R., ROSS, A.H., TAKAGAKI, Y. and KHORANA, H.G. (1981) Proc. Natl. Acad. Sci U.S.A. *78,* 2225.

BENESKI, D.A. and CATTERALL, W.A. (1980) Proc. Natl. Acad. Sci. U.S.A. *77,* 639.

BENISEK, W.F. (1977) Methods Enzymol. *46,* 469.

BERCOVICI, T., GITLER, C. and BROMBERG, A. (1978) Biochemistry *17,* 1484.

BERHANU, P., OLESKY, J.M., TSAI, P., THAMM, P., SAUNDERS, D. and BRANDENBURG, D. (1982) Proc. Natl. Acad. Sci. U.S.A. *79,* 4069.

BERNIER, M. and ESCHER, E. (1980) Helv. Chim. Acta *63,* 1308.

BISHOP, W.H., HENKE, L., CHRISTOPHER, J.P. and MILLAR, D.B. (1980) Proc. Natl. Acad. Sci. U.S.A. *77,* 1980.

BISPINK, L. and MATTHAEI, J.H. (1973) FEBS Lett. *37*, 291.

BISPINK, L. and MATTHAEI, H. (1977) Methods Enzymol. *46*, 637.

BISSON, R. and MONTECUCCO, C. (1981) Biochem. J. *193*, 757.

BISSON, R., AZZI, A., GUTWENIGER, H., COLONNA, R., MONTECUCCO, C. and ZANOTTI, A. (1978) J. Biol. Chem. *253*, 1874.

BISSON, R., MONTECUCCO, C. and CAPALDI, R.A. (1979) FEBS Lett. *106*, 317.

BISSON, R., JACOBS, B. and CAPALDI, R.A. (1980) Biochemistry *19*, 4173.

BRADLEY, G.F., EVANS, W.B.L. and STEVENS, I.D.R. (1977) J. Chem. Soc. Perkin Trans. *2*, 1214.

BRANDT, J., FREDRIKSSON, M. and ANDERSSON, L.-O. (1974) Biochemistry *13*, 4758.

BREMS, D.N. and RILLING, H.C. (1979) Biochemistry *18*, 860.

BREMS, D.N., BRUENGER, E. and RILLING, H.C. (1981) Biochemistry *20*, 3711.

BRESLOW, R., FEIRING, A. and HERMAN, F. (1974) J. Am. Chem. Soc. *96*, 5937.

BRIDGES, A.J. and KNOWLES, J.R. (1974) Biochem. J. *143*, 663.

BROWNE, D.J., HIXSON, S.S. and WESTHEIMER, F.H. (1971) J. Biol. Chem. *246*, 4477.

BRUNNER, J. (1981) Trends Biochem. Sci. *6*, 44.

BRUNNER, J. and RICHARDS, F.M. (1980) J. Biol. Chem. *255*, 3319.

BRUNNER, J. and SEMENZA, G. (1981) Biochemistry *20*, 7174.

BRUNNER, J., SENN, H. and RICHARDS, F.M. (1980) J. Biol. Chem. *255*, 3313.

BRUNSWICK, D.J. and COOPERMAN, B.S. (1971) Proc. Natl. Acad. Sci. U.S.A. *68*, 1801.

BRUNSWICK, D.J. and COOPERMAN, B.S. (1973) Biochemistry *12*, 4074.

BUDKER, V.G., KNORRE, D.G., KRAVCHENKO, V.V., LAVRIK, O.I., NEVINSKY, G.A. and TEPLOVA, N.H. (1974) FEBS Lett. *49*, 159.

BURNETTE, W.N. (1981) Analyt. Biochem. *112*, 195.

CALVERT, J.G. and PITTS, J.N. (1966) Photochemistry, Wiley, New York.

CANOVA-DAVIS, E. and RAMACHANDRAN, J. (1980) Biochemistry *19*, 3275.

CANTOR, C.R. (1980) Ann. N.Y. Acad. Sci. *346*, 379.

CAO, T.M. and SUNG, M.T. (1982) Biochemistry *21*, 3419.

CARLIER, M.-F., HOLOWKA, D.A. and HAMMES, G.G. (1979) Biochemistry *18*, 3452.

CARNEY, D.H., GLENN, K.C., CUNNINGHAM, D.D., DAS, M., FOX, C.F. and FENTON, J.W. (1979) J. Biol. Chem. *254*, 6244.

CARTWRIGHT, I.L. and HUTCHINSON, D.W. (1980) Nucleic Acids Res. *8*, 1675.

CARTWRIGHT, I.L., HUTCHINSON, D.W. and ARMSTRONG, V.W. (1976) Nucleic Acids Res. *3*, 2331.

CASNELLIE, J.E., SCHLICHTER, D.J., WALTER, U. and GREENGARD, P. (1978) J. Biol. Chem. *253*, 4771.

CERLETTI, N. and SCHATZ, G. (1979) J. Biol. Chem. *254*, 7746–7751.

CHAKRABARTI, P. and KHORANA, H.G. (1975) Biochemistry *14*, 5021.

CHAPMAN, O.L. (1979) Pure Appl. Chem. *51*, 331.

CHATTERJEE, P.K. and CANTOR, C.R. (1978) Nucleic Acids Res. *5*, 3619.

CHLADEK, S., QUIGGLE, K., CHINALI, G., KOHUT, J. and OFENGAND, J. (1977) Biochemistry *16*, 4312.

CHEN, M.S., CHANG, P.K. and PRUSSOFF, W.H. (1976) J. Biol. Chem. *251*, 6555.

CHEN, S. and GUILLORY, R.J. (1979) J. Biol. Chem. *254,* 7220.

CHEN, S. and GUILLORY, R.J. (1981) J. Biol. Chem. *256,* 8318.

CHICHEPORTICHE, R., BALERNA, M., LOMBET, A., ROMEY, G. and LAZDUNSKI, M. (1979) J. Biol. Chem. *254,* 1552.

CHONG, P.C.S. and HODGES, R.S. (1981) J. Biol. Chem. *256,* 5064.

CHOWDHRY, V. and WESTHEIMER, F.H. (1978a) J. Am. Chem. Soc. *100,* 309.

CHOWDHRY, V. and WESTHEIMER, F.H. (1978b) Biorganic Chem. *7,* 189.

CHOWDHRY, V. and WESTHEIMER, F.H. (1979) Ann. Rev. Biochem. *48,* 293.

CHOWDHRY, V., VAUGHAN, R., and WESTHEIMER, F.H. (1976) Proc. Natl. Acad. Sci. U.S.A. *73,* 1406.

COLMAN, R., SCRIVEN, E.F.V., SUSCHITSKY, H. and THOMAS, D.R. (1981) Chem. Ind. p. 249.

COMENS, P.G., SIMMER, R.L. and BAKER, J.B. (1982) J. Biol. Chem. *257,* 42.

COOPERMAN, B.S. and BRUNSWICK, D.J. (1973) Biochemistry *12,* 4079.

COOPERMAN, B.S., JAYNES, E.N., BRUNSWICK, D.J. and LUDDY, M.A. (1975) Proc. Natl. Acad. Sci. U.S.A. *72,* 2974.

CORNELISSE, J. and HAVINGA, E. (1975) Chem. Rev. *75,* 353.

CORNELISSE, J., DEGUNST, P. and HAVINGA, E. (1975) Adv. Phys. Org. Chem. *11,* 225.

CORNELISSE, J., LODDER, G. and HAVINGA, E. (1979) Rev. Chem. Intermed. *2,* 231.

CZARNIECKI, M.F. and BRESLOW, R. (1979) J. Am. Chem. Soc. *101,* 3675.

CZARNECKI, J., GEAHLEN, R. and HALEY, B. (1979) Methods Enzymol. *56,* 642.

DARFLER, F.V. and MARINETTI, G.V. (1977) Biochem. Biophys. Res. Commun. *79,* 1.

DAS, M. and FOX, C.F. (1978) Proc. Natl. Acad. Sci. U.S.A. *75,* 2644.

DAS, M. and FOX, C.F. (1979) Ann. Rev. Biophys. Bioeng. *8,* 165.

DAS, M., MIYAKAWA, T., FOX, C.F., PRUSS, R.M., AHARONOV, A. and HERSCHMAN, H.R. (1977) Proc. Natl. Acad. Sci. U.S.A. *74,* 2790.

DEGRAFF, B.A., GILLESPIE, D.W. and SUNDBERG, R.J. (1974) J. Am. Chem. Soc. *96,* 7491.

DELPIERRE, G.R. and FRUTON, J.S. (1965) Proc. Natl. Acad. Sci. U.S.A. *54,* 1161.

DE LUCA, N., BZIK, D., PERSON, S. and SNIPES, W. (1981) Proc. Natl. Acad. Sci. U.S.A. *78,* 912.

DEMOLIOU, C.D. and EPAND, R.M. (1980) Biochemistry *19,* 4539.

DEMOLIOU-MASON, C. and EPAND, R.M. (1982) Biochemistry *21,* 1989.

DE RIEMER, L.H. and MEARES, C.F. (1981) Biochemistry *20,* 1612.

DOCKTER, M.E. (1979) J. Biol. Chem. *254,* 2161.

DOERING, W.v.E., BUTTERY, R.G., LAUGHLIN, R.G. and CHAUDHURI, N. (1956) J. Am. Chem. Soc. *78,* 3224.

DRAPER, M.W., NISSENSON, R.A., WINER, J., RAMACHANDRAN, J. and ARMAND, C.D. (1982) J. Biol. Chem. *257,* 3714.

DREYFUSS, G., SCHWARTZ, K., BLOUT, E.R., BARRIO, J.R., LIU, F.T. and LEONARD, N.J. (1978) Proc. Natl. Acad. Sci. *75,* 1199.

DURE, L.S., SCHRADER, W.T. and O'MALLEY, B.W. (1980) Nature (London) *283,* 784.

EBERLE, A. and SCHWYZER, R. (1976) Helv. Chim. Acta *59,* 2421.

EID, P., GOELDNER, M.P., HIRTH, C.G. and JOST, P. (1981) Biochemistry *20,* 2251.

EISENTHAL, K.B., TURRO, N.J., AIKAWA, M., BUTCHER, J.A., DU PUY, C., HEFFERON, G., HETHERINGTON, W., KORENOWSKI, G.M. and MCAULIFFE, M.J. (1980) J. Am. Chem. Soc. *102,* 6563.

ENOCH, H.G., FLEMING, P.J. and STRITTMATTER (1979) J. Biol. Chem. *254,* 6483.

ERECINSKA, M. (1977) Biochem. Biophys. Res. Commun. *76,* 495.

ERECINSKA, M., VANDERKOOI, J.M. and WILSON, D.F. (1975) Arch. Biochem. Biophys. *171,* 108.

ERNI, B. and KHORANA, H.G. (1980) J. Am. Chem. Soc. *102,* 3888.

ESCHER, E. (1977) Helv. Chim. Acta *60,* 339.

ESCHER, E. and SCHWYZER, R. (1974) FEBS Lett. *46,* 347.

ESCHER, E. and SCHWYZER, R. (1975) Helv. Chim. Acta *58,* 1465.

ESCHER, E.H.F., NGUYEN, T.M.D., GUILLEMETTE, G. and REGOLI, D.C. (1978) Nature (London) *275,* 145.

ESCHER, E.H.F., NGUYEN, T.M.D., ROBERT, H., ST.-PIERRE, S.A. and REGOLI, D.C. (1978) J. Med. Chem. *21,* 860.

ESCHER, E.H.F., ROBERT, H. and GUILLEMETTE, G. (1979) Helv. Chim. Acta *62,* 1217.

EVANS, E.A. (1976) Self-Decomposition of Radiochemicals: Principles, Control Observations and Effects, The Radiochemical Centre, Amersham, England.

FAHRENHOLZ, F. and SCHIMMACK, G. (1975) Hoppe-Seyler's Z. Physiol. Chem. *356,* 469.

FANNIN, F.F., EVANS, J.O., GIBBS, E.M. and DIEDRICH, D.F. (1981) Biochim. Biophys. Acta *649,* 189.

FARLEY, R.A., GOLDMAN, D.W. and BAYLEY, H. (1980) J. Biol. Chem. *255,* 860.

FISCHLI, W., CAVIEZEL, M., EBERLE, A., ESCHER, E. and SCHWYZER, R. (1976) Helv. Chim. Acta *59,* 878.

FISER, I., SCHEIT, K.H., STOFFLER, G. and KUECHLER, E. (1974) Biochem. Biophys. Res. Commun. *60,* 1112.

FISHER, C.E. and PRESS, E.M. (1974) Biochem. J. *139,* 135.

FLEET, G.W.J., PORTER, R.R. and KNOWLES, J.R. (1969) Nature *224,* 511.

FLEET, G.W.J., KNOWLES, J.R. and PORTER, R.R. (1972) Biochem. J. *128,* 499.

FORBUSH, B. and HOFFMAN, J.F. (1979) Biochim. Biophys. Acta *555,* 299.

FORBUSH, B., KAPLAN, J. and HOFFMAN, J.F. (1978) Biochemistry *17,* 3667.

GALARDY, R.E., CRAIG, L.C. and PRINTZ, M.P. (1973) Nature (London) New Biol. *242,* 127.

GALARDY, R.E., CRAIG, L.C., JAMIESON, J.D. and PRINTZ, M.P. (1974) J. Biol. Chem. *249,* 3510.

GALARDY, R.E., HULL, B.E. and JAMIESON, J.D. (1980) J. Biol. Chem. *255,* 3148.

GANJIAN, I., PETTEI, M.J., NAKANISHI, K. and KAISSLING, K.E. (1978) Nature (London) *271,* 157.

GEAHLEN, R.L. and HALEY, B.E. (1979) J. Biol. Chem. *254,* 11982.

GEAHLEN, R.L., HALEY, B.E. and KREBS, E.G. (1979) Proc. Natl. Acad. Sci. U.S.A. *76,* 2213.

GIRDLESTONE, J., BISSON, R. and CAPALDI, R.A. (1981) Biochemistry *20,* 152.

GIRSHOVICH, A.S., BOCHKAREVA, E.S., KRAMAROV, V.M. and OVCHINNIKOV, Y.A. (1974) FEBS Lett. *45,* 213.

GIRSHOVICH, A.S., POZDNYAKOV, V.A. and OVCHINNIKOV, Y.A. (1976) Eur. J. Biochem. *69,* 321.

GLAZER, A.N. (1976) The Proteins *2,* 1.

GLAZER, A.N., DELANGE, R.J. and SIGMAN, D.S. (1975) In: Laboratory Techniques in Biochemistry and Molecular Biology, Vol. 4, Part I. Work, T.S. and Work E., eds., North-Holland, Amsterdam.

GOELDNER, M.P. and HIRTH, C.G. (1980) Proc. Natl. Acad. Sci. U.S.A. *77,* 6439.

GOELDNER, M.P., HIRTH, C.G., KIEFFER, B. and OURISSON, G. (1982) Trends. Biochem. Sci. 7, 310.

GOLDMAN, P.W., POBER, J.S., WHITE, J. and BAYLEY, H. (1979) Nature (London) 280, 841.

GOLDSTEIN, J.A., MCKENNA, C. and WESTHEIMER, F.H. (1976) J. Am. Chem. Soc. *98,* 7327.

GORMAN, J.J. and FOLK, J.E. (1980) J. Biol. Chem. *255,* 1175.

GOTTSCHALK, M.E. and KEMP, R.G. (1981) Biochemistry *20,* 2245.

GRANT, P.G., STRYCHARZ, W.A., JAYNES, E.N. and COOPERMAN, B.S. (1979) Biochemistry *18,* 2149.

GREENBERG, G.R., CHAKRABARTI, P. and KHORANA, H.G. (1976) Proc. Natl. Acad. Sci. U.S.A. *73,* 86.

GREENWALL, P., JEWETT, S.L. and STARK, G.R. (1973) J. Biol. Chem. *248,* 5994.

GRIMSHAW., J. and DESILVA, A.P. (1981) Chem. Soc. Rev. *10,* 181.

GRONEMEYER, H. and PONGS, O. (1980) Proc. Natl. Acad. Sci. U.S.A. *77,* 2108.

GUPTA, C.M., COSTELLO, C.E. and KHORANA, H.G. (1979) Proc. Natl. Acad. Sci. U.S.A. *76,* 3139.

GUPTA, C.M., RADHAKRISHNAN, R., GERBER, G.E., OLSEN, W.L., QUAY, S.C. and KHORANA, H.G. (1979) Proc. Natl. Acad. Sci. U.S.A. *76,* 2595.

GUTHROW, C.E., RASMUSSEN, H. BRUNSWICK, D.J. and COOPERMAN, B.S. (1973) Proc. Natl. Acad. Sci. U.S.A. *70,* 3344.

HALL, C. and RUOHO, A. (1980) Proc. Natl. Acad. Sci. U.S.A. *77,* 4529.

HANSTEIN, W.G., HATEFI, Y. and KIEFER, H. (1979) Biochemistry *18*, 1019.

HAVINGA, E. and CORNELISSE, J. (1976) Pure Appl. Chem. *47,* 1.

HAVRON, A. and SPERLING, J. (1977) Biochemistry *16,* 5631.

HEARST, J.E. (1981) J. Invest. Dermatol. *77,* 39.

HENDERSON, R., JUBB, J.S. and WHYTOCK, S. (1978) J. Mol. Biol. *123,* 259.

HENKEN, J. (1977) J. Biol. Chem. *252,* 4293.

HEXTER, C.S. and WESTHEIMER, F.H. (1971a) J. Biol. Chem. *246,* 3934.

HEXTER, C.S. and WESTHEIMER, F.H. (1971b) J. Biol. Chem. *246,* 3928.

HINDS, T.R. and ANDREASEN, T.J. (1981) J. Biol. Chem. *256,* 7877.

HIXSON, S.S. and HIXSON, S.H. (1973) Photochem. Photobiol. *18,* 135.

HIXSON, S.H. and HIXSON, S.S. (1975) Biochemistry *14,* 4251.

HOYER, P.B., OWENS, J.R. and HALEY, B.E. (1980) Ann. N.Y. Acad. Sci. *346,* 280.

HSIUNG, N. and CANTOR, C.R. (1974) Nucleic Acids Res. *1*, 1753.

HSIUNG, N., REINES, S.A. and CANTOR, C.R. (1974) J. Mol. Biol. *88*, 841.

HUANG, C.K. and RICHARDS, F.M. (1977) J. Biol. Chem. *252*, 5514.

HUANG, K-S., RADHAKRISHNAN, R., BAYLEY, H. and KHORANA, H.G. (1982) J. Biol. Chem. *257*, 13616.

HU, V.W. and WISNIESKI, B.J. (1979) Proc. Natl. Acad. Sci. U.S.A. *76*, 5460-5464.

HUTCHINSON, D.W. and MUTOPO, D.S. (1979) Biochem. J. *181*, 779.

IDDON, B., METH-COHN, O., SCRIVEN, E.F.V., SUSCHITSKY, H. and GALLAGHER, P.T. (1979) Angew. Chem. Int. Ed. Engl. *18*, 900.

ISAACS, S.T., SHEN, C.-K. J., HEARST, J.E. and RAPOPORT, H. (1977) Biochemistry *16*, 1058-1064.

ISAEV, S.D., YURCHENKO, A.G., STEPANOV, F.N., KOLYADA, G.G., NOVIKOV, S.S., and KARPENKO, N.F. (1973) Zh. Org. Khim. *9*, 724.

IVANOVSKAYA, M.G., SOKOLOVA, N.I. and SHABAROVA, Z.A. (1979) Biorg. Khim. *5*, 35.

JACOBS, S., HAZUM, E., SHECHTER, Y. and CUATRECASAS, P. (1979) Proc. Natl. Acad. Sci. U.S.A. *76*, 4918.

JAFFE, C.L., LIS, H. and SHARON, N. (1979) Biochem. Biophys. Res. Commun. *91*, 402.

JAFFE, C.L., LIS, H. and SHARON, N. (1980) Biochemistry *19*, 4423.

JAKOBY, W.B. and WILCHEK, M. (1977) Methods Enzymol. 46.

JELENC, P.C., CANTOR, C.R. and SIMON, S.R. (1978) Proc. Natl. Acad. Sci. U.S.A. *75*, 3564.

JENG, S.J. and GUILLORY, R.J. (1975) J. Supramol. Struct. *3*, 448.

JI, T.H. (1977) J. Biol. Chem. *252*, 1566.

JI, T.H. (1979) Biochim. Biophys. Acta *559*, 39.

JI, T.H., KIEHM, D.J. and MIDDAUGH, C.R. (1980) J. Biol. Chem. *255*, 2990.

JI, I. and JI, T.H. (1981) Proc. Natl. Acad. Sci. U.S.A. *78*, 5465.

JOHNSON, G.L., MACANDREW, V.I. and PILCH, P.F. (1981) Proc. Natl. Acad. Sci. U.S.A. *78*, 875.

JØRGENSEN, P.L., KARLISH, S.J.D. and GITLER, C. (1982) J. Biol. Chem. *257*, 7435.

KACZOROWSKI, G.J., LE BLANC, G. and KABACK, H.R. (1980) Proc. Natl. Acad. Sci. U.S.A. *77*, 6319.

KAHANE, I. and GITLER, C. (1978) Science *201*, 351.

KARLISH, S.J.D., JØRGENSEN, P.L. and GITLER, C. (1977) Nature (London) *269*, 715.

KATZENELLENBOGEN, J.A. and HSIUNG, H.M. (1975) Biochemistry *14*, 1736.

KATZENELLENBOGEN, J.A., MYERS, H.N. and JOHNSON, H.J. (1973) J. Org. Chem. *38*, 3525.

KATZENELLENBOGEN, J.A., JOHNSON, H.J., CARLSON, K.E. and MYERS, H.N. (1974) Biochemistry *13*, 2986.

KATZENELLENBOGEN, J.A., MEYERS, H.N., JOHNSON, H.J., KEMPTON, R.J. and CARLSON, K.E. (1977) Biochemistry *16*, 1964.

KELLEY, J.A., NAU, N., FORSTER, H.-J. and BIEMANN, K. (1975) Biomed. Mass Spectrom. *2*, 313.

KERLAVAGE, A.R. and TAYLOR, S.S. (1980) J. Biol. Chem. *255*, 8483.

KERR, J.A., O'GRADY, B.V. and TROTMAN-DICKENSON, A.F. (1967) J. Chem. Soc. A, 897.
KIEHM, D.J. and JI, T.H. (1977) J. Biol. Chem. *252,* 8524.
KIRMSE, W. (1964) Carbene Chemistry, Academic Press, New York.
KLAUSNER, Y.S., FEIGENBAUM, A.M., DE GROOT, N. and HOCHBERG, A.A. (1978) Arch. Biophys. Biochem. *185,* 151.
KLEIN, G., SATRE, M., DIANOUX, A.-C. and VIGNAIS, P.V. (1981) Biochemistry *20,* 1339.
KLIP, A. and GITLER, C. (1974) Biochem. Biophys. Res. Commun. *60,* 1155.
KNAUF, P.A., BREUER, W., MCCULLOCH, L. and ROTHSTEIN, A. (1978) J. Gen. Physiol. *72,* 631.
KNOWLES, J.R. (1972) Acc. Chem. Res. *5,* 155.
KUPFER, A., GANI, V., JIMENEZ, J.S. and SHALTIEL, S. (1979) Proc. Natl. Acad. Sci. U.S.A. *76,* 3073.
KYBA, E.P. and ABRAMOVITCH, R.A. (1980) J. Am. Chem. Soc. *102,* 735.
LASCU, I., KEZDI, M., GOIA, I., JEBELEANU, G., BARZU, O., PANSINI, A., PAPA, S. and MANTSCH, H.H. (1979) Biochemistry *18,* 4818.
LAU, E.P., HALEY, B.E. and BARDEN, R.E. (1977a) Biochemistry *16,* 2581.
LAU, E.P., HALEY, B.E. and BARDEN, R.E. (1977b) Biochem. Biophys. Res. Commun. *76,* 843.
LAUQUIN, G., POUGEOIS, R. and VIGNAIS, P.V. (1980) Biochemistry *19,* 4620.
LAWSON, W.B. and SCHRAMM, H.J. (1962) J. Am. Chem. Soc. *84,* 2017.
LEE, T.T., WILLIAMS, R.E. and FOX, C.F. (1979) J. Biol. Chem. *254,* 11787.
LEHMAN, P.A. and BERRY, R.S. (1973) J. Am. Chem. Soc. *95,* 8614.
LEPOCK, J.R., THOMPSON, J.E., KRUUV, J. and WALLACH, D.F.H. (1978) Biochem. Biophys. Res. Commun. *85.* 344.
LEVY, D. (1973) Biochim. Biophys. Acta *322,* 329.
LEWIS, R.V. and ALLISON, W.S. (1978) Arch. Biophys. Biochem. *190,* 163.
LEWIS, R.V., ROBERTS, M.F., DENNIS, E.A. and ALLISON, W.S. (1977) Biochemistry *16,* 5650.
LIFTER, J., HEW, C.-L., YOSHIOKA, M., RICHARDS, F.F. and KONIGSBERG, W.H. (1974) Biochemistry *13,* 3567.
LIN, S.-Y. and RIGGS, A.D. (1974) Proc. Natl. Acad. Sci. U.S.A. *71,* 947.
LINDEMANN, J.G. and LOVINS, R.E. (1976) Analyt. Biochem. *75,* 682.
LINSLEY, P.S., BLIFELD, C., WRANN, M. and FOX, C.F. (1979) Nature (London) *278,* 745.
LOMANT, A.J. and FAIRBANKS, G., (1976) J. Mol. Biol. *104,* 243.
LOUVARD, D., SEMERIVA, M. and MAROUX, S. (1976) J. Mol. Biol. *106,* 1023.
LUTTER, L.C. (1982) J. Biol. Chem. 257, 1577.
LWOWSKI, W., ed. (1970) Nitrenes, Wiley, New York.
LWOWSKI, W. (1980) Ann. N.Y. Acad. Sci. *346,* 491.
MAASSEN, J.A. (1979) Biochemistry *18,* 1288.
MAASSEN, J.A. and MOLLER, W. (1974) Proc. Natl. Acad. Sci. U.S.A. *71,* 1277.
MAASSEN, J.A. and MOLLER, W. (1978) J. Biol. Chem. *253,* 2777.
MACFARLANE, D.E., MILLS, D.C.B. and SRIVASTAVA, P.C. (1982) Biochemistry *21,* 544.

MALY, P., RINKE, J., ULMER, E., ZWIEB, C. and BRIMACOMBE, R. (1980) Biochemistry *19*, 4179.

MARKWELL, M.A.K. and FOX, C.F. (1980) J. Virol. 33, 152.

MARINETTI, T.D., OKAMURA, M.Y. and FEHER, G. (1979) Biochemistry *18*, 3126.

MARTYR, R.J. and BENISEK, W.F. (1973) Biochemistry *12*, 2172.

MARTYR, R.J. and BENISEK, W.F. (1975) J. Biol. Chem. *250*, 1218.

MAS, M.T., WANG, J.K. and HARGRAVE, P.A. (1980) Biochemistry *19*, 684.

MATHESON, R.R. and SCHERAGA, H.A. (1979) Biochemistry *18*, 2437.

MATHESON, R.R., VAN WART, H.E., BURGESS, A.W., WEINSTEIN, L.I. and SCHERAGA, H.A. (1977) Biochemistry *16*, 396.

MATZKE, A.J.M., BARTA, A. and KUECHLER, E. (1980) Proc. Natl. Acad. Sci. U.S.A. *77*, 5110.

MCROBBIE, I.M., METH-COHN, O. and SUSCHITZKY, H. (1976) Tetrahedron Lett. 925.

MEANS, G.E. and FEENEY, R.E. (1971) Chemical Modification of Proteins, Holden-Day Inc., San Francisco.

MIKKELSEN, R.B. and WALLACH, D.F.H. (1976) J. Biol. Chem. *251*, 7413.

MOHLER, H., BATTERSBY, M.K. and RICHARDS, J.G. (1980) Proc. Natl. Acad. Sci. U.S.A. *77*, 1666.

MOHLER, H., RICHARDS, J.G. and WU, J.-Y. (1981) Proc. Natl. Acad. Sci. U.S.A. *78*, 1935.

MONTECUCCO, C., BISSON, R., GACHE, C. and JOHANNSSON, A. (1981) FEBS Lett. *128*. 17.

MORELAND, R.B. and DOCKTER, M.E. (1980) Analyt. Biochem. *103*, 26.

MORELAND, R.B. and DOCKTER, M.E. (1981) Biochem. Biophys. Res. Commun *99*, 339.

MORELAND, R.B., SMITH, P.K., FUJIMOTO, E.K. and DOCKTER, M.E. (1982) Analyt. Biochem. *121*, 321.

MOSS, R.A. (1980) Acc. Chem. Res. *13*, 58.

MOSS, R.A. and JONES, M., eds. (1973) Carbenes, Vol. I, Wiley, New York.

MOSS, R.A. and JONES, M., eds. (1975) Carbenes, Voll. II, Wiley, New York.

MOSS, R. and JONES, M. (1978) Reactive Intermediates *1*, 69.

MOSS, R.A., FEDORYNSKI, M. and SHIEH, W.-C. (1979) J. Am. Chem. Soc. *101*, 4736.

MUNSON, K.-B. and KYTE, J. (1981) J. Biol. Chem. *256*, 3223.

MURACHI, T. and YASUI, M. (1965) Biochemistry *4*, 2275.

MURAMOTO, K. and RAMACHANDRAN, J. (1980) Biochemistry *19*, 3280.

MURAMOTO, K. and RAMACHANDRAN, J. (1981a) Biochemistry *20*, 3376.

MURAMOTO, K. and RAMACHANDRAN, J. (1981b) Biochemistry *20*, 3380.

MUROV, S.L. (1973) Handbook of Photochemistry, Dekker, New York.

MURRAY, A. and LLOYD WILLIAMS, D. (1958) Organic Syntheses With Isotopes, Parts I and II, Wiley, New York.

NAKAYAMA, H., NOZAWA, M. and KANAOKA, Y. (1979) Chem. Pharm. Bull. *27*, 2775.

NATHANSON, N.M. and HALL, Z.W. (1980) J. Biol. Chem. *255*, 1698.

NEWMAN, J., FOSTER, D.L., WILSON, T.H. and KABACK, H.R. (1981) J. Biol. Chem. *256*, 11804.

NGO, T.T., YAM, C.F., LENHOFF, H.M. and IVY, J. (1981) J. Biol. Chem. *256*, 11313.

NIEDEL, J., DAVIS, J. and CUATRECASAS, P. (1980) J. Biol. Chem. *255,* 7063.

NIELSEN, P.E. and BUCHARDT, O. (1982) Photochem. Photobiol. *35,* 317.

NIELSEN, P.E., LEICK, V. and BUCHARDT, O. (1978) FEBS Lett. *128,* 17.

NOEL, D., NIKAIDO, K. and AMES, G.F. (1979) Biochemistry *18,* 4159.

NORDEEN, S.K., LAU, N.C., SHOWERS, M.O. and BAXTER, J.D. (1981) J. Biol. Chem. *256,* 10503.

O'FARRELL, P.H. (1975) J. Biol. Chem. *250,* 4007.

OFENGAND, J., GORNICKI, P., CHAKRABURTTY, K. and NURSE, K. (1982) Proc. Natl. Acad. Sci. U.S.A. *79,* 2817.

OGEZ, J.R., TIVOL, W.F. and BENISEK, W.F. (1977) J. Biol. Chem. *252,* 6151.

OLSON, H. MCK., GRANT, P.G., COOPERMAN, B.S. and GLITZ, D.G. (1982) J. Biol. Chem. *257,* 2649.

OSTE, C., PARFAIT, R., BOLLEN, A. and CRICHTON, R.R. (1977) Mol. Gen. Genet. *152,* 253.

OWEN, M.J., KNOTT, J.C.A. and CRUMPTON, M.J. (1980) Biochemistry *19,* 3092–3099.

PACKMAN, L.C. and PERHAM, R. (1982) Biochemistry *21,* 5171.

PARK, C.S., HILLEL, Z. and WU, C.-W. (1982a) J. Biol. Chem. *257,* 6944.

PARK, C.S., WU, F.Y.-H. and WU, C.-W. (1982b) J. Biol. Chem. *257,* 6950.

PARSONS, B.J. (1980) Photochem. Photobiol. *32,* 813.

PAYNE, D.W., KATZENELLENBOGEN, J.A. and CARLSON, K.E. (1980) J. Biol. Chem. *255,* 10359.

PETERS, K. and RICHARDS, F.M. (1977) Ann. Rev. Biochem. *46,* 523.

PFEUFFER, T. (1977) J. Biol. Chem. *252,* 7224.

PFISTER, K., STEINBACK, K.E., GARDNER, G. and ARNTZEN, C.J. (1981) Proc. Natl. Acad. Sci. U.S.A. *78,* 981.

POLITZ, S.M., NOLLER, H.F. and MCWHIRTER, P.D. (1981) Biochemistry *20,* 372.

POMERANTZ, A.H., RUDOLPH, S.A., HALEY, B.E. and GREENGARD, P. (1975) Biochemistry *14,* 3858.

POTTER, R. and HALEY, B. (1983) Methods Enzymol. *91*, 613.

PROCHASKA, L., BISSON, R. and CAPALDI, R.A. (1980) Biochemistry *19,* 3174.

QUAY, S.C., RADHAKRISHNAN, R. and KHORANA, H.G. (1981) J. Biol. Chem. *256,* 4444.

RADHAKRISHNAN, R., GUPTA, C.M., ERNI, B., ROBSON, R.J., CURATOLO, W., MAJUMDAR, A., ROSS, A.H., TAKAGAKI, Y. and KHORANA, H.G. (1980) Ann. N.Y. Acad. Sci. *346,* 165.

RADHAKRISHNAN, R., ROBSON, R.J., TAKAGAKI, Y. and KHORANA,H.G. (1981) Methods Enzymol. *72D,* 408.

RADHAKRISHNAN, R., COSTELLO, C.E. and KHORANA, H.G. (1982) J. Am. Chem. Soc. *104,* 3990.

RAMACHANDRAN, J.; HAGMAN, J. and MURAMOTO, K. (1981) J. Biol. Chem. *256,* 11424.

RANDOLPH, L.V. and ALLISON, W.S. (1978) Arch. Biochem. Biophys. *190,* 163.

RANGEL-ALDAO, R., KUPIEC, J.W. and ROSEN, O.M. (1979) J. Biol. Chem. *254,* 2499.

RASHIDBAIGI, A. and RUOHO, A.E. (1981) Proc. Natl. Acad. Sci. *78,* 1609.

REISER, A. and LEYSHON, L.J. (1970) J. Am. Chem. Soc. *92,* 7487.

REISER, A. and LEYSHON, L.J. (1971) J. Am. Chem. Soc. *93,* 4051.

REISER, A., TERRY, G.C. and WILLETS, F.N. (1966) Nature (London) *211,* 410.

REISER, A., WILLETS, F.W., TERRY, G.C., WILLIAMS, V. and MARLEY, R. (1968) Trans. Faraday Soc. *64,* 3265.

RICHARDS, F.F., LIFTER, J., HEW, C.-L., YOSHIOKA, M. and KONIGSBERG, W.H. (1974) Biochemistry *13,* 3572.

RICHARDS, F.M. and VITHAYATHIL, P.J. (1960) Brookhaven Symp. Biol. *13,* 115.

RICHARDS, F.M. and BRUNNER, J. (1980) Ann. N.Y. Acad. Sci. *346,* 144.

RINKE, J., MEINKE, M., BRIMACOMBE, R., FINK, G., ROMMEL, W. and FASOLD, H. (1980) J. Mol. Biol. *137,* 301.

ROBSON, R.J., RADHAKRISHNAN, R., ROSS, A.H., TAKAGAKI, Y. and KHORANA, H.G. (1982) In: Lipid–Protein Interactions, Vol. 2, Jost, P.C. and GRIFFITHS, O.H. eds., Wiley, New York.

ROSS, A.H., RADHAKRISHNAN, R., ROBSON, R.J. and KHORANA, H.G. (1982) J. Biol. Chem. *257,* 4152.

RUOHO, A.E., KIEFER, H., ROEDER, P. and SINGER, S.J. (1973) Proc. Natl. Acad. Sci. U.S.A. *70,* 2567.

RUOHO, A. and KYTE, J. (1974) Proc. Natl. Acad. Sci. U.S.A. *71,* 2352.

RUOHO, A. and KYTE, J. (1977) Methods Enzymol. *46,* 523.

SADLER, S.E. and MALLER, J.L. (1982) J. Biol. Chem. *257,* 355.

SATOR, V., GONZALEZ-ROS, J.M., CALVO-FERNANDEZ, P. and MARTINEZ-CARRION, M. (1979) Biochemistry *18,* 1200.

SAWADA, F. and KANBAYASHI, N. (1973) J. Biochem. *74,* 459.

SCHALTMANN, K. and PONGS, O. (1982) Proc. Natl. Acad. Sci. U.S.A. *79,* 6.

SCHMITZ, E. and OHME, R. (1961) Chem. Ber. *94,* 2166.

SCHOELLMAN, G. and SHAW, E. (1963) Biochemistry *2,* 252.

SCHWARTZ, I. and OFENGAND, J. (1978) Biochemistry *17,* 2524.

SCHWARTZ, M.A., DAS, O.P. and HYNES, R.O. (1982) J. Biol. Chem. *257,* 2343.

SCHWARTZ, D.C., SAFFRAN, W., WELSH, J., HAAS, R., GOLDENBERG, M. and CANTOR, C.R. (1982) Cold Spring Harbor Symp. Quant. Biol. *47*, 189.

SCHWYZER, R. and CALVIEZEL, M. (1971) Helv. Chim. Acta *54,* 1395.

SCRIVEN, E.F.V., ed. (1984) Azides and Nitrenes, Academic Press, New York.

SEBALD, W., MACHLEIDT, W. and WACHTER, E. (1980) Proc. Natl. Acad. Sci. U.S.A. *77,* 785.

SEELA, F. (1976) Z. Naturforschung C *31,* 389.

SEELA, F. and ROSEMEYER, H. (1977) Hoppe Seyler's Z. Physiol. Chem. *358,* 129.

SEN, R. (1982) Ph. D. Thesis, Columbia University.

SEN, R., CARRIKER, J.D., BALOGH-NAIR, V. and NAKANISHI, K. (1982) J. Am. Chem. Soc. *104,* 3214.

SHAFER, J., BARONOWSKY, P., LAURSEN, R., FINN, F. and WESTHEIMER, F.H. (1966) J. Biol. Chem. *241,* 421.

SHAMA, R.K. and KHARASH, N. (1968) Angew. Chem. Int. Ed. *7,* 37.

SHARW, E. (1970a) The Enzymes *1,* 91.
SHAW, E. (1970b) Physiol. Rev. *50,* 244.
SHETLAR, M.D. (1980) Photochem. Photobiol. Rev. *5,* 105.
SHEPPARD, G. (1972) Atomic Energy Rev. *10,* 3.
SHORR, R.G.L., HEALD, S.R., JEFFS, P.W. LAVIN, T.N., STROHSACKER, M.W., LEFKOWITZ, R.J. and CARON, M.G. (1982) Proc. Natl. Acad. Sci. U.S.A. *79,* 2778.
SIGRIST, H. and ZAHLER, P. (1982) In: Membranes and Transport, Vol. 1. A. Martonosi, ed. Plenum, New York, pp. 173-184.
SILMAN, I. and KARLIN, A. (1969) Science *164,* 1420.
SINGER, S.J. (1967) Adv. Prot. Chem. *22,* 1.
SINGH, A., THORNTON, E.R. and WESTHEIMER, P.C. (1962) J. Biol. Chem. *237,* 3006.
SKARE, K., BLACK, J.L., PANCOE, W.L. and HALEY, B. (1977) Arch. Biochem. Biophys. *180,* 409.
SMITH, D.P., KILBOURN, M.R., MCDOWELL, J.H. and HARGRAVE, P.A. (1981) Biochemistry *20,* 2417.
SMITH, K.C., ed. (1976) Aging, Carcinogenesis, and Radiation Biology, Plenum, New York.
SMITH, P.A.S. and BROWN, B.B. (1951) J. Am. Chem. Soc. *73,* 2435.
SMITH, P.A.S. and BOYER, J.H. (1963) Org. Synth. *4,* 75.
SMITH, R.A.G. and KNOWLES, J.R. (1973) J. Am. Chem. Soc. *95,* 5072.
SMITH, R.A.G. and KNOWLES, J.R. (1974) Biochem. J. *141,* 51.
SMITH, R.A.G. and KNOWLES, J.R. (1975) J. Chem. Soc. Perkin Trans. *2,* 686.
SMITH, S.B. and BENISEK, W.F. (1980) J. Biol. Chem. *255,* 2690.
SMOLINSKY, G., WASSERMAN, E. and YAGER, W.A. (1962) J. Am. Chem. Soc. *84,* 3220.
SONENBERG, N., ZAMIR, A. and WILCHEK, M. (1974) Biochem. Biophys. Res. Commun. *59,* 693.
SONG, P.-S. and TAPLEY, K.J. (1979) Photochem. Photobiol. *29,* 1177.
SPIESS, M., BRUNNER, J. and SEMENZA, G. (1982) J. Biol. Chem. *257,* 2370.
STACKHOUSE, J. and WESTHEIMER, F.H. (1981) J. Org. Chem. *46,* 1891.
STADEL, J.M., GOODMAN, D.B.P., GALARDY, R.E. and RASMUSSEN, H. (1978) Biochemistry *17*, 1403.
STANDRING, D.N. and KNOWLES, J.R. (1980) Biochemistry *19,* 2811.
STAROS, J.V. (1980) Trends Biochem. Sci. *5,* 320.
STAROS, J.V. and KNOWLES, J.R. (1978) Biochemistry *17,* 3321.
STAROS, J.V. and RICHARDS, F.M. (1974) Biochemistry *13,* 2720.
STAROS, J.V. HALEY, B.E. and RICHARDS, F.M. (1974) J. Biol. Chem. *249,* 5004.
STAROS, J.V., RICHARDS, F.M. and HALEY, B.E. (1975) J. Biol. Chem. *250,* 8174.
STAROS, J.V., BAYLEY, H., STANDRING, D.N. and KNOWLES, J.R. (1978) Biochem. Biophys. Res. Commun. *80,* 568.
STAROS, J.V., MORGAN, D.G. and APPLING, P.R. (1981) J. Biol. Chem. *256,* 5890.
STECK, T.L. and YU, J. (1973) J. Supramol. Struct. *1,* 220.
STOFFEL, W. and METZ, P. (1982) Hoppe-Seyler's Z. Physiol. Chem. *363,* 19.
STOFFEL, W., SCHREIBER, C. and SCHEEFERS, H. (1978) Hoppe-Seyler's Z. Physiol. Chem.

359, 923.

STOFFEL, W., SALM, K.-P. and MULLER, M. (1982) Hoppe-Seyler's Z. Physiol. Chem. *363,* 1.

SUNDBERG, R.J., GILLESPIE, D.W. and DEGRAFF, B.A. (1975) J. Am. Chem. Soc. *97,* 6193.

SUTOH, K. (1980) Biochemistry *19,* 1977.

SWANSON, R.A. and DUS, K.M. (1979) J. Biol. Chem. *254,* 7238.

TAKAGAKI, Y., GUPHA, C. and KHORANA, H.G. (1980) Biochem. Biophys. Res. Commun. *95,* 589.

TAKAHASHI, K. (1968) J. Biol. Chem. *243,* 6171.

TAKEUCHI, H. and KOYAMA, K. (1981) Chem. Comm. p. 202.

TARRAB-HAZDAI, R., BERCOVICI, T., GOLDFARB, V. and GITLER, C. (1980) J. Biol. Chem. *255,* 1204.

TAYLOR, C.A., SMITH, H.E. and DANZO, B.J. (1980) Proc. Natl. Acad. Sci. U.S.A. *77,* 234.

THAMM, P., SAUNDERS, D. and BRANDENBURG, D. (1980) In: Insulin, Brandenburg, D. and Wollmer, A., eds., de Gruyter, Berlin, pp. 309–316.

THOMAS, J.W. and TALLMAN, J.F. (1981) J. Biol. Chem. *256,* 9838.

TURRO, N.J. (1979) Modern Molecular Photochemistry, Benjamin/Cummings, Menlo Park, California.

TURRO, N.J., BUTCHER, J.A., MOSS, R.A., GUO, W., MUNJAL, R.C. and FEDORYNSKI, M. (1980) J. Am. Chem. Soc. *102,* 7576.

VAN DER WALT, B., NIKODEM, V.M. and CAHNMANN, H.J. (1982) Proc. Natl. Acad. Sci. U.S.A. *79,* 3508.

VANDEST, P., LABBE, J.-P. and KASSAB, R. (1980) Eur. J. Biochem. *104,* 433.

VANIN, E.F. and JI, T.H. (1981) Biochemistry 20, 6754.

VAUGHAN, R.J. and WESTHEIMER, F.H. (1969a) Analyt. Biochem. *29,* 305.

VAUGHAN, R.J. and WESTHEIMER, F.H. (1969b) J. Am. Chem. Soc. *91,* 217.

VAUGHAN, R. (1970) Ph. D. Thesis, Harvard University.

VAVER, V.A., USHAKOV, A.N. and TSYRENINA, M.L. (1979) Bioorg. Khim. *5,* 1520.

WALLACE, L.J. and FRAZIER, W.A. (1979a) Proc. Natl. Acad. Sci. U.S.A. *76,* 4250.

WALLACE, L.J. and FRAZIER, W.A. (1979b) J. Biol. Chem. *254,* 10109.

WALTER, U., UNO, I., LIU, A.Y.-C. and GREENGARD, P. (1977) J. Biol. Chem. *252,* 6494.

WANG, K. and RICHARDS, F.M. (1974) J. Biol. Chem. *249,* 8005.

WANG, S.Y., ed. (1976) Photochemistry and Photobiology of Nucleic Acids, Vols. I and II, Academic Press, New York.

WELLS, E. and FINDLAY, J.B.C. (1980) Biochem. J. *187,* 719.

WENTRUP, C. (1979) Reaktive Zwischenstufen, Vol. I, Thieme Stuttgart.

WESTHEIMER, F.H. (1980) Ann. N.Y. Acad. Sci. *346,* 134.

WICKNER, W.T. (1976) Proc. Natl. Acad. Sci. U.S.A. *73,* 1159.

WILLIAMS, N. and COLEMAN, P.S. (1982) J. Biol. Chem. *257,* 2834.

WILSON, D.F., MUKAI, Y., ERECINSKA, M. and VANDERKOOI, J.M. (1975) Arch. Biochem. Biophys. *171,* 104.

WISHER, M.H., BARON, M.D., JONES, R.H., SONKSEN, P.H., SAUNDERS, D.J., THAMM, P.

and BRANDENBURG, D. (1980) Biochem. Biophys. Res. Commun. *92,* 492.
WISNIESKI, B.J. and BRAMHALL, J.S. (1981) Nature (London) *289,* 319.
WISNIESKI, B.J. and IWATA, K.K. (1977) Biochemistry *16,* 1321.
WITZEMANN, M. and RAFTERY, M.A. (1978) Biochemistry *17,* 3598.
WITZEMANN, V., MUCHMORE, D. and RAFTERY, M.A. (1979) Biochemistry *18,* 5511.
WOLLENZIEN, P.L. and CANTOR, C.R. (1982) Proc. Natl. Acad. Sci. U.S.A. *79,* 3940.
WONG, P.C., GRILLER, D. and SCAIANO, J.C. (1981) J. Am. Chem. Soc. *103,* 5935.
WRENN, S.M. and HOMCY, C.J. (1980) Proc. Natl. Acad. Sci. U.S.A. *77,* 4449.
YEUNG, C.W.T., MOULE, M.L. and YIP, C.C. (1980) Biochemistry *19,* 2196.
YIP, C.C., YEUNG, C.W.T. and MOULE, M.L. (1978) J. Biol. Chem. *253,* 1743.
YIP, C.C., YEUNG, C.W.T. and MOULE, M.L. (1980) Biochemistry *19,* 70.
YOSHIOKA, M., LIFTER, J., HEW, C.-L., CONVERSE, C.A., ARMSTRONG, M.Y.K., KONIGSBERG, W.H. and RICHARDS, F.F. (1973) Biochemistry *12,* 4679.
YUE, V.T. and SCHIMMEL, P.R. (1977) Biochemistry *16,* 4678.
ZUPANCIC, J.J. and SCHUSTER, G.B. (1980) J. Am. Chem. Soc. *102,* 5958.
ZUPANCIC, J.J., GRASSE, P.B. and SCHUSTER, G.B. (1981) J. Am. Chem. Soc. *103,* 2423.

Subject index

Where specific compounds and macromolecules are indexed they are either photochemical reagents or their targets. For example, chymotrypsin is not indexed where it has been used as a reagent for cleaving polypeptide chains, but, it is indexed where attempts have been made to define the substrate binding site by photoaffinity labeling.